Sukhoi Interceptors

Yefim Gordon

Original translation by Dmitriy Komissarov

MIDLAND
An imprint of
Ian Allan Publishing

Sukhoi Interceptors
© 2004 Yefim Gordon

ISBN 1 85780 180 6

Published by Midland Publishing
4 Watling Drive, Hinckley, LE10 3EY, England
Tel: 01455 254 490   Fax: 01455 254 495
E-mail: midlandbooks@compuserve.com

Midland Publishing is an imprint of
Ian Allan Publishing Ltd

Worldwide distribution (except North America):
Midland Counties Publications
4 Watling Drive, Hinckley, LE10 3EY, England
Telephone: 01455 254 450   Fax: 01455 233 737
E-mail: midlandbooks@compuserve.com
www.midlandcountiessuperstore.com

North American trade distribution:
Specialty Press Publishers & Wholesalers Inc.
39966 Grand Avenue, North Branch, MN 55056, USA
Tel: 651 277 1400   Fax: 651 277 1203
Toll free telephone: 800 895 4585
www.specialtypress.com

© 2004 Midland Publishing
Design concept and layout
by Polygon Press Ltd. (Moscow, Russia)
Line drawings by V. I. Klimov and Polygon-Press

This book is illustrated with photos by Yefim Gordon and Sergey Popsuyevich, as well as from the archives of the Sukhoi OKB, Yefim Gordon, the Flight Research Institute, Wojskowa Agencja Fotograficzna and *Aviatsiya i Vremya* magazine

Printed in England by Ian Allan Printing Ltd
Riverdene Business Park, Molesey Road,
Hersham, Surrey, KT12 4RG

All rights reserved. No part of this publication may be reproduced, stored in a retrieval system, transmitted in any form or by any means, electronic, mechanical or photo-copied, recorded or otherwise, without the written permission of the publishers.

Contents

Introduction 3
1. The Second Generation. 5
2. The Flying Pipes 27
3. Defending the Homeland:
   The Su-9/Su-11 in Service 33
4. Experimental Models. 43
5. Sukhoi Strikes Back 49
6. 'Target in Sight!': Su-15 in Action . . . 81
7. A Close Look at the Sukhoi Deltas . . 89
   Line drawings 104
   Colour Photographs 114

Title page: A Su-15TM banks away from the camera, displaying the two missiles (R-98MT, port; R-98MR, starboard) and UPK-23-250 cannon pods. A similarly configured aircraft shot down Korean Air Lines Boeing 747-230B HL7442 on the night of 1st September 1983.

Below: PVO pilots wearing pressure suits and full-face pressure helmets climb into two late-production Su-15TMs featuring four missile pylons each. Note the MiG-25s in the background.

Front cover: '25 Blue' (c/n 0306), the sole prototype of the Su-15*bis*.
Rear cover, top: Su-11 '14 Blue' (c/n 0115307) at the Central Russian Air Force Museum; bottom: Su-9 '07 Blue' in the base museum at Savasleyka AB.

Introduction

I'm Back!

By the early 1950s the Moscow-based OKB-155 (*opytno-konstrooktorskoye byuro* – experimental design bureau; the number is a code allocated for security reasons) headed by Artyom Ivanovich Mikoyan had effectively monopolised fighter design in the Soviet Union. OKB-115 was losing its glamour after its head Aleksandr Sergeyevich Yakovlev had lost his position in the government as Vice People's Commissar (ie, Vice-Minister) of the Aircraft Industry; Semyon Alekseyevich Lavochkin's OKB-301 was being gradually reoriented from aircraft to missile design by the Powers That Be. OKB-134 headed by Chief Designer Pavel Osipovich Sukhoi fared even worse: having developed and tested the Su-9, Su-11 and Su-15 twinjet fighters (the first aircraft to bear these designations), the design bureau was liquidated in 1949 – officially with the crash of the Su-15 prototype as the pretext, though other reasons were involved as well. The last two aircraft developed by the OKB prior to the closure – the Su-17 fighter (again the first to be thus designated) and the Su-10 four-turbojet medium bomber – never had a chance to prove their worth and the prototypes were scrapped without ever being flown.

The world-famous Mikoyan/Gurevich MiG-15 was produced by eight (!) aircraft factories in the Soviet Union alone, not to mention foreign production, becoming the main (or *unified*, in the terminology of the time) jet fighter of the Soviet Air Force – or, if we are to call a spade a spade, its sole tactical fighter type; it was soon succeeded by its more refined derivative, the MiG-17. The *izdeliye* SM series of fighter prototypes entered test in 1952, culminating in the SM-9 – the prototype of the Soviet Union's first truly supersonic fighter, the MiG-19, and OKB-155 had a host of other projects in the making. (*Izdeliye* (product) such and such was a common code for Soviet military hardware items.) It appeared that nobody could challenge the positions of A. I. Mikoyan and his closest aide Mikhail Iosifovich Gurevich.

Yet the year 1953 changed things dramatically – not just because the adored and feared Soviet leader Iosif V. Stalin died, ending a whole era in the nation's history. That year major changes took place in the Soviet defence industry in general and the aircraft industry in particular. Among other things, the design bureaux of Pavel O. Sukhoi and Vladimir Mikhaïlovich Myasishchev were re-established – and it was just as well that they were. Both OKBs went on to create outstanding examples of Soviet aviation technology which not only occupied an important place in Soviet/Russian military aviation but won international acclaim.

Sukhoi's new assignment as head of OKB-1 in May 1953 was a particularly unexpected move. Nobody in the nation's aircraft industry took this seriously: everybody thought that the bosses of the People's Commissariat of Aircraft Industry (NKAP – *Narodnyy komissariaht aviatsionnoy promyshlennosti*) were simply pursuing their own ends by unseating one chief designer who had fallen from favour and replacing him with another man who was temporarily back in from the cold. They were quite wrong.

At that time the Korean War was still raging. This conflict showed that five years after the Second World War the good old dogfight tactics were still good, despite the fact that propeller-driven fighters had been replaced by transonic jets (and they would still be good ten years from then, as the Vietnam War would demonstrate!). The ongoing quest for speed and altitude begun in the 1920s

Pavel Osipovich Sukhoi, Hero of the Soviet Union, suffered the ignominy of having his first OKB closed in 1949 but was vindicated in 1953 when the Sukhoi OKB was re-established.

Above: The first aircraft to be designated Su-9 was this experimental fighter bearing the manufacturer's designation *izdeliye* K. The similarity to the Messerschmitt 262 is obvious but this is not a copy.

Above: The 'first-generation' Su-11 (*izdeliye* LK) was a version of the 'Su-9 Mk I' powered by Lyul'ka TR-1 engines instead of RD-10s (Junkers Jumo 004Bs), hence the L. It likewise remained in prototype form.

The 'first-generation' Su-15 was an unusual fighter, with the two engines in a staggered-tandem arrangement and cockpit offset to port. A radar was to be housed in the bullet fairing above the air intake.

brought about a massive improvement in the performance of all classes of aircraft. In the case of bombers, flying higher and faster was the only means of evading enemy fighters and anti-aircraft defences. The 'fighter makers' followed suit, knowing that flying higher and faster was required to get at the escaping enemy.

That said, Sukhoi was absolutely dissatisfied with the project inherited from OKB-1's former head V. V. Kondrat'yev – namely reverse-engineering the North American F-86A Sabre, an example of which had fallen into Soviet hands in Korea. Hence, winning support from what was now the Ministry of Aircraft Industry (MAP – *Ministerstvo aviatsionnoy promyshlennosti*), in August 1953 Sukhoi managed to get the Soviet Council of Ministers to issue a directive authorising him to develop aircraft of his own.

In November the reborn OKB finally received its own premises – a section of OKB-155 located on the south side of Moscow's Central Airfield named after Mikhail V. Frunze (aka Khodynka), a mere 6 km (3.7 miles) from the Kremlin. Before the war the premises had been occupied by OKB-51 led by the famous 'fighter king' Nikolay Nikolayevich Polikarpov; after his death and the demise of his OKB the place had been home to a missile systems design bureau under Vladimir N. Chelomey in 1944-52.

On 15th January 1954 Pavel O. Sukhoi's OKB-1 and its prototype construction facility were renumbered, inheriting the number of the Polikarpov OKB. Now they were officially designated the State Experimental Factory No.51; hereinafter it will be called OKB-51, since this designation used in numerous official documents became the 'last name' of virtually all Sukhoi aircraft which later won fame. Ye. S. Fel'sner was appointed Chief Designer Sukhoi's depute, later joined in this capacity as N. G. Zyrin and V. A. Alybin; they were responsible for the powerplant, airframe and systems/equipment respectively.

In the early 1950s the use of delta wings utilising a thin airfoil and a high wing loading was accepted as the way to achieve high speeds; yet such wings incurred a marked deterioration of the fighter's manoeuvrability and field performance. As a result, close-in dogfighting was gradually replaced by missile attacks at long range as the main tactic. Later, however, high agility and built-in cannons made a comeback.

For now, however, the wing design was the greatest problem area in the development of many advanced combat aircraft. A heated argument went on at the Soviet Union's top authority on aircraft design – the Central Aero- & Hydrodynamics Institute named after Nikolay Ye. Zhukovskiy (TsAGI – *Tsentrahl'nyy aero- i ghidrodinamicheskiy instiktoot*) – as to what wing planform was best for aircraft designed for high supersonic speeds. Sometimes the researchers even disbelieved their own findings. Former OKB-1 employee Yevgeniy Adler (who went on to work for OKB-51) helped resolve the issue, cutting away the trailing edge of the rhomboid wings proposed by TsAGI and obtaining delta wings. As a matter of fact, OKB-1 had placed its bets on delta wings back when Kondrat'yev was the boss.

Only practice could determine whether swept wings or delta wings were best. Working in both directions, OKB-51 proposed to the Soviet Air Force that its aircraft projects featuring delta wings would have designations with the letter T (for *treugol'noye krylo*), while those incorporating swept wings would have the letter S (*strelovidnoye krylo*).

A further note must be made on the designations of Sukhoi aircraft. It is a fact that the aircraft designed by Sukhoi's former OKB were designated consecutively Su-1 through Su-17. The first aircraft developed by the new OKB-51 received the service designation Su-7 which had been used for a piston-engined design of Second World War vintage; quite simply, this stood for 'Sukhoi aircraft with a Lyul'ka AL-7F engine'. The next design of OKB-51 became the Su-9, and from then on Sukhoi aircraft received odd numbers, even numbers being allocated only as an exception.

Chapter 1
The Second Generation

S-1 and T-1 tactical fighters
S-3 and T-3 interceptors (project stage)

After its resurrection in 1953 the Sukhoi OKB (OKB-51) started work in two main areas, developing a tactical fighter and a dedicated interceptor for the Air Defence Force. A draft directive 'On the development of new fast tactical fighters with swept and delta wings' prepared by the Soviet Council of Ministers said, '2. The Ministry of Defence Industry (D. F. Ustinov) and Chief Designer P. O. Sukhoi are hereby authorised to design and build a single-seat experimental fighter with delta wings and a turbojet engine designed by A. M. Lyul'ka for the purpose of further enhancing the performance [of fighters] and mastering the new layout of fighters.' After the principal performance parameters had been set at a session of the Board of the Ministry of Aircraft Industry, on 5th August 1953 the Soviet Council of Ministers issued directive No.2072-839 ordering the development of these two aircraft. According to this document each aircraft was to be developed in two very different versions, one featuring swept wings and the other delta wings.

Shortly afterwards the Soviet Air Force (VVS – *Voyenno-vozdooshnyye seely*) presented a general operational requirement (GOR) for the two aircraft. The interceptor version was to have a maximum speed of 1,900-1,950 km/h (1,180-1,210 mph) and a service ceiling of 19,000-20,000 m (62,335-65,620 ft), climbing to 15,000 m (49,210 ft) in two minutes flat. Effective range at 10,000 m (32,800 ft) was to be 1,400 km (870 miles) in 'clean' configuration (ie, without external stores) and 2,250 km (1,397 miles) with two drop tanks. The interceptor was to be armed with two 30-mm (1.18 calibre) cannons.

By mid-1954 OKB-51 had prepared an advanced development project (ADP) and built a full-scale mock-up of the swept-wing interceptor which received the in-house designation S-3; however, further work in this direction was deemed inexpedient by MAP and the S-3 was abandoned. By autumn the OKB had completed a joint ADP for the delta-wing T-1 fighter and T-3 interceptor. These passed the internal review stage and then the so-called mock-up review commission (a project analysis by the customer for the purpose of eliminating grave shortcomings at an early stage) with flying colours in October-November, whereupon both types were cleared for prototype construction. Interestingly, originally the T-1 was originally allocated higher priority, the T-3 being relegated to second place.

To speed up development the T-3 featured considerable commonality both with the T-1 and with the swept-wing S-1 tactical fighter which was by then at a more advanced development stage (this aircraft, which eventually emerged as the famous Su-7 fighter-bomber, lies outside the scope of this book). All three aircraft were to be powered by a single AL-7 afterburning turbojet developed by Arkhip Mikhaïlovich Lyul'ka's OKB-165; known in-house as *izdeliye* 21, this engine offered a thrust of 7,500 kgp (16,530 lbst) at full military power and 10,000 kgp (22,045 lbst) in full afterburner. The monocoque fuselage had a circular cross-section and featured a nose air intake (a pitot-type intake) divided by a vertical splitter into two ducts flanking the cockpit.

A special research team headed by P. P. Krasil'shchikov was formed at TsAGI to explore the aerodynamics of Sukhoi's future fighters. The team recommended that a TsAGI S9 series symmetrical airfoil with a thickness/chord ratio of about 6% and a rounded leading edge be used for the wings of the T-1 and T-3. TsAGI suggested that all-movable tailplanes (slab stabilisers, or stabilators) be used on the T-1/T-3; similar recommendations were made to OKB-155 with regard to the SM-9 (the future MiG-19), Ye-2 and Ye-4 fighters.

The cockpit featured a bubble canopy with an aft-sliding hood. A curious feature of the projected T-1/T-3 was that in the event of ejection, clamps on the ejection seat headrest engaged lugs on the sliding portion of the canopy so that the latter stayed with the seat, protecting the pilot against the slipstream. (The same principle was used by the Mikoyan OKB for the Ye-2, Ye-4 and Ye-7 prototypes, the latter of which evolved into the production MiG-21F; however, these aircraft had a one-piece forward-hinged canopy.) The control system utilised irreversible hydraulic actuators in all three control circuits.

The T-1 and T-3 differed primarily in forward fuselage design. Like the S-1, the T-1 fighter had an axisymmetrical air intake with a movable conical centrebody (shock cone); the latter was to accommodate an SRD-3 Grad (Hail; pronounced *grahd*) gun ranging radar, a reverse-engineered version of the American AN/APG-30 fitted to the F-86A Sabre (SRD = *samolyotnyy rahdiodal'nomer* – aircraft-mounted radio rangefinder). Conversely, the T-3 was to feature an indigenous Almaz-3 (Diamond-3) fire control radar developed by OKB-15, a branch of NII-17 headed by Viktor V. Tikhomirov (the latter establishment became the Moscow Research Institute of Instrument Engineering (MNIIP – *Moskovskiy naoochno-issledovatel'skiy institoot priborostroyeniya*), aka NPO Vega-M, in 1967). The Almaz was a twin-antenna radar, with separate search and target tracking antennas. This necessitated the use of a fixed-area air intake, the search antenna being located in a small conical radome on the upper lip and the tracking antenna in a similarly sized hemispherical radome on the air intake splitter. The aerodynamics of supersonic flight were not yet properly studied in the Soviet Union at the time, hence little effort was made to maximise inlet efficiency.

The two aircraft also differed in armament fit: the T-1 was to have three 30-mm (1.18 calibre) cannons buried in the wing roots (one to port and two to starboard) whereas the T-3 had only one cannon on each side. The ammunition supply in both cases was 65 rounds per gun. Both types had provisions for carrying pods with 57-mm (2.24-in.) ARS-57 *Skvorets* (Starling) folding-fin aircraft rockets (FFARs) on underwing pylons.

Detail design of the T-1 was completed in December 1954, construction of the prototype and a static test airframe at OKB-51's experimental production facility in Moscow (MMZ No.51) commencing a while earlier (in November). The work on the T-3 took rather longer, the detail design stage being completed in May 1955 and prototype construction beginning in April. In the course of detail design the T-3 underwent a major change, switching from a monocoque fuselage structure to a semi-monocoque fuselage with longerons. Another important change was the incorporation of a third fuel tank in the rear fuselage beneath the engine to provide the required range in the event that T-2 or T-4 jet fuel grades (which were lighter than the T-1 aviation kerosene considered initially) were used; this increased total fuel capacity to 3,180 litres (699.6 Imp gal) and the fuel load to 2,600 kg (5,730 lb). Two 500-litre (110 Imp gal) drop tanks could be carried side by side on pylons under the centre fuselage, giving an extra 820 kg (1,810 lb) of fuel.

Another important change concerned the armament. A Council of Ministers directive issued on 30th December 1954 amended the specific operational requirement (SOR) for the T-3 to the effect that the interceptor was to be armed with air-to-air missiles – specifically, the K-7L AAM under development at OKB-134 under I. I. Toropov or the K-6V AAM being developed by Pavel D. Grooshin's OKB-2.

In those days each Soviet aircraft design bureau developed its own ejection seats for its combat aircraft. Thus OKB-51 created the KS-1 seat for the S-1 and T-1/T-3 (KS = [*katapool'tnoye*] *kres*lo *Sukhovo* – Sukhoi [ejection] seat). Starting in April 1955, the KS-1 underwent trials on one of three UTI-MiG-15 trainers converted into ST-10 ejection seat testbeds by the Flight Research Institute named after Mikhail M. Gromov (LII – *Lyotno-issledovatel'skiy institoot*); the aircraft in question was most probably serialled '401ᵁ Blue' (ie, 401ᵞ in Cyrillic characters). The experimental seat was installed in the rear cockpit which had the standard sliding canopy section replaced by a longer canopy designed for the Sukhoi jets. However, the first ejections with dummies showed that the arrangement initially chosen with the canopy doubling as a slipstream shield incurred major technical problems which proved difficult to overcome. Hence the engineers soon opted for a more traditional arrangement where the canopy was jettisoned prior to ejection. The new arrangement underwent further testing on the ST-10 and a suitably modified Yakovlev Yak-25 interceptor known as the Yak-25L, another LII testbed, in 1956, allowing the S-1 and T-3 prototypes to be equipped with prototype KS-1 seats in a timely manner.

T-3 experimental interceptor (*izdeliye* 81)

By May 1955 MAP had shifted its priorities, attaching greater importance to the T-3 delta-wing interceptor; a similar situation existed at the Mikoyan OKB which was instructed by MAP to concentrate on interceptors. By the end of the year the T-1 tactical fighter programme had been scrapped altogether, leaving only the S-1 in the running in this area. However, as already mentioned, construction of the T-1 prototype had begun in November 1954, and by the end of 1955 the work was well advanced. This was when the commonality between the T-1 and T-3 proved an asset. Not wishing to let the effort be wasted, the Sukhoi OKB decided to convert the airframe into the flying prototype of the T-3 (known in-house as *izdeliye* 81). The conversion involved replacing the fuselage nose up to and including the cockpit section, as well as changes to the forward bays of the wings.

By the end of 1955 the static test example of the T-3 had been completed; so had all major airframe subassemblies of the flying prototype except the wings, and overall programme readiness for the first flight had reached 95%. Due to the later delivery of some equipment items the prototype lacked the Almaz-3 radar, the PVU-67 computing sight (*pritsel'no-vychislitel'noye oostroystvo*) designed for aiming the K-7L and K-6V AAMs, and the SRZO-2 identification friend-or-foe interrogator/transponder (*samolyotnyy rahdiolokatsionnyy zaproschik-otvetchik* – aircraft-mounted radar [IFF] interrogator/responder). Instead, the avionics bay in the fuselage nose was equipped by test equipment and ballast to maintain the correct centre of gravity position. Nor were the cannons fitted at this stage. This was no great problem – after all, nobody expects a first prototype to be fully combat-capable. Much the worse for wear, engine development was running behind schedule, as OKB-165 had run into problems with the AL-7F and could not complete the bench tests in time; also, it transpired that the engine was rather heavier than anticipated.

The Sukhoi OKB's experimental shop completed the airframe of the first prototype T-3 (construction number 01?) in March 1956, and a flight-cleared prototype engine was delivered to MMZ No.51 for installation in the aircraft in early April. On the night of 23rd April the aircraft was trucked to LII's airfield in the town of Zhukovskiy south of Moscow where OKB-51 had its flight test facility, like nearly all other Soviet aircraft design bureaux. V. N. Makhalin, a pilot seconded to the Sukhoi OKB from the Soviet Air Force State Research Institute named after Valeriy P. Chkalov (GK NII VVS – *Gosoodarstvennyy krasnoznamyonnyy naoochno-issledovatel'skiy institoot voyenno-vozdooshnykh seel*) a short while earlier, was appointed project test pilot by a special MAP order, with M. I. Zooyev as engineer in charge of the flight tests.

Throughout the following month the aircraft underwent ground systems checks and taxying tests. Finally, on 26th April 1956, after a lengthy delay caused by a malfunctioning communications radio that took some time to fix, the T-3 prototype took to the air with

The T-3 prototype during trials. Note the metal 'radome' tipped by an air data boom and the gun blast plates ahead of the wing roots (no cannons are fitted). The K-7L missiles are carried for aerodynamic testing only, since no radar is fitted.

Top: This head-on view illustrates the T-3's characteristic intake design; the upper radome is for the search antenna and the lower one for the guidance antenna.
Centre and above: Two views of the T-3 with the pylons removed. Note the tail bumper and the cooling air intakes just aft of the wing trailing edge.

Three views of the T-3 during its public debut at Moscow-Tushino on 24th June 1956. These views illustrate the additional pitots on the wings and the anti-flutter booms on the stabilators.

Makhalin at the controls. Less than a month later, on 24th June, the aircraft made its public debut, participating in the annual Aviation Day flypast at Moscow's Tushino airfield in company with another brand-new Sukhoi aircraft, the S-1 fighter-bomber prototype. Interestingly, Western aviation experts immediately and unerringly recognised both aircraft as being Sukhoi products; in many other cases their guesses as to the origin of new Soviet combat aircraft had been wide of the mark. The T-3 and S-1 exhibited certain 'family traits' that were to become Sukhoi hallmarks for the next 20 years, including characteristically graceful tails topped by green dielectric fairings. After the T-3's Tushino debut the NATO's Air Standards Co-ordinating Committee (ASCC) allocated the reporting name *Fishpot* to the new interceptor; this was subsequently amended to *Fishpot-A* when the production versions became known to the West.

The manufacturer's flight tests proceeded in several stages, the aircraft being grounded from time to time by the need to make various modifications and engine changes. Stage A lasted until 28th September and was concerned with exploring the T-3's flight envelope; in included 31 flights, 27 of which were test flights. At this stage the interceptor's performance with K-7L missiles was determined, stability and handling at high angles of attack were checked, even spinning trials performed. On 1st September V. N. Makhalin who performed these trials gained the distinction of being the first Sukhoi test pilot to make a dead-stick landing in a supersonic jet fighter – especially a delta-wing aircraft. When a spin was initiated at 10,000 m (32,800 ft), a compressor stall occurred and the engine flamed out. Recovering from the spin, the pilot put the aircraft into a shallow dive and set about restarting the engine. This proved to be no easy thing; he managed it on the fifth try, making a safe landing.

By late October the original engine had reached the limit of its service life and the T-3 was sent back to the OKB's experimental shop to have a new one fitted. The aircraft was in lay-up until early March 1957; apart from the engine change, it underwent a number of modifications. The intended Almaz-3 radar and PVU-67 computing sight were installed, as was a missile launch control system for the K-7L AAMs and the SRZO-2 IFF. A new *Mindahl'* (Almond) communications radio was fitted instead of the original RSIU-4 Doob (Oak) radio, a new sliding canopy portion was installed and the brake parachute container was enlarged.

Stage B of the manufacturer's flight tests began considerably behind schedule due to the late delivery of the replacement engine. At this stage, which began on 8th March 1957,

LII test pilots V. M. Pronyakin and Vladimir S. Ilyushin (the latter was later transferred to the Sukhoi OKB) started flying the T-3. In the summer of 1957 the aircraft was ferried to the Ministry of Defence's 6th Test Range at Vladimirovka AB near Akhtoobinsk (Saratov Region) located in the delta of the Volga River; this base would later become the main facility of GK NII VVS. There between 1st June and 23rd August the K-7L missile system was verified, which included live missile launches, and the engine in-flight restarting procedure was verified, among other things; Vladimir S. Ilyushin flew the aircraft at this stage.

On 28th August the T-3 returned to Zhukovskiy. After nearly a month's lull the tests resumed on 20th September and were completed on 16th October 1957; the vibration characteristics of the K-7L missile/pylon combination and the missile-armed aircraft as a whole were determined at the closing stage. Apart from the pilots already mentioned, test pilots L. G. Kobishchan, A. A. Koznov and M. L. Petushkov also flew the T-3 for familiarisation purposes.

All in all, in the course of 18 months the T-3 had made nearly 80 flights totalling 38 hours 21 minutes; the results obtained in these flights confirmed that the chosen layout was sound and the basic design features worked. The aircraft showed quite decent performance, even though not all of the target figures had been met. This was because the gross weight was much higher than the project figure, while the engine performance figures advertised by the manufacturer proved decidedly optimistic. As a result, range and endurance fell short of the expectations, while the unstick and landing speed were higher than anticipated.

The following is an excerpt from the manufacturer's flight test report of 'the T-3 experimental interceptor No.01' (sic – ie, c/n 01?).

'1. The manufacturer's flight tests revealed that the aircraft possesses the following performance:

a) top speed in level flight in full afterburner at 12,000 m [39,370 ft] – 2,100 km/h [1,304 mph];

b) climb time to 10,000 m [32,800 ft] from start of stable climb from 1,000 m [3,280 ft] at full military power – 2.3 minutes;

c) service ceiling in afterburner mode – 18,000 m [59,055 ft];

d) technical range at 12,000 m without drop tanks – 1,440 km [894 miles]; with drop tanks, 1,840 km [1,140 miles];

e) endurance at 12,000 m without drop tanks – 1 hour 39 minutes; with drop tanks, 2 hours 10 minutes.

2. The K-7L [missile] carriage and launch system operates faultlessly throughout the explored range of speeds (Mach numbers) and altitudes.

Another in-flight view of the T-3, showing details of the underside and the 'grinning' air intake. Note the different shades of skin on the wings.

K-7L missiles were fired at altitudes of 5,100-18,300 m [16,730-60,040 ft]. The missile launch does not affect engine operation and does not require the use of a fuel check valve (throttling back the engine automatically to prevent surge caused by missile exhaust gas ingestion – Auth.).

The aircraft's handling and manoeuvrability ensure normal flight operation.

In order to facilitate piloting during future operation of the T-3 it is necessary:

1. to replace the control cables in the aileron and tailplane control circuits with rigid linkages;

2. to install an ARZ-1 artificial-feel unit adjusting stick forces, depending on the flight mode;

3. to install an AP-106 [yaw] damper in the directional control circuit.

4. The landing gear, hydraulics, powerplant and other systems of the aircraft have been brought up to an adequate reliability level in the course of the manufacturer's flight tests and permit safe operation of the aircraft.'

It may be added that the take-off run was 1,050-1,150 m (3,445-3,770 ft) and the landing run without the use of a brake parachute was 1,780-1,840 m (5,840-6,040 ft).

The greatest problems encountered during the T-3's tests were caused by the AL-7F engine which was still suffering from teething troubles and was extremely capricious. Suffice it to say that in the course of the 38-hour flight test programme the aircraft had to undergo four engine changes! The AL-7F ran at a high temperature, necessitating the provision of additional engine cooling air scoops on the centre fuselage and replacement of wiring bundles in the engine bay with heat-resistant ones.

Building on experience gained with the S-1 which was powered by the same engine, the T-3's never-exceed speed was limited to Mach 1.83 for the duration of the manufacturer's flight tests in order to avoid engine surge. Another remedy was to incorporate so-called bleed bands (ie, bleed valves) at the AL-7F's fourth and fifth compressor stages. The downside of this feature was that the valves markedly reduced available thrust at speeds in excess of Mach 1.6; instead of 7,500 kgp (16,530 lbst) at full military power and 10,000 kgp (22,045 lbst) in full afterburner the modified engine delivered only 6,850 kgp (15,100 lbst) and 8,950 kgp (19,730 lbst) respectively. Hence, when powered by an AL-7F incorporating bleed valves, the T-3 could not do better than 1,830 km/h (1,136 mph) and 18,000 m (59,055 ft); the 2,100 km/h top speed quoted in the manufacturer's flight test report applied to a configuration with an engine lacking bleed valves.

Test pilots noted an excessively sharp reaction to aileron inputs; this led to the recommendation to incorporate push-pull rods in the pitch and roll control circuits and provide an artificial-feel unit and a yaw damper. Originally the T-3 featured perforated airbrakes for maximum braking efficiency; however, the pilots reported that the airbrakes created an infernal sound when deployed and the perforated airbrakes were replaced with 'solid' ones.

Also, it became clear that the aircraft would be unable to reach the specified maximum speed and altitude with a fixed-area air intake and the existing radome arrangement; by then, however, OKB-51 was working on more efficient intake designs. The following is a brief structural description of the T-3.

Type: Single-engined single-seat supersonic interceptor designed for day and night operation in visual meteorological conditions (VMC) and instrument meteorological conditions (IMC). The airframe is of all-metal construction. Length overall 17.07 m (56 ft 0 in), wing span 8.54 m (28 ft 0¼ in), height on ground 4.82 m (15 ft 9¾ in), wing area 34.0 m² (365.5 sq.ft), flap area 3.515 m² (37.79 sq.ft), aileron area 1.73 m² (18.6 sq.ft), tailplane area 5.58 m² (60.0 sq.ft),

Fuselage: Semi-monocoque riveted stressed-skin structure of circular cross-section Structurally the fuselage consists of two sections: forward (section F-1) and rear (section F-2), the latter being detachable for engine maintenance and removal.

The *forward fuselage* incorporates a fixed-area nose air intake divided by a vertical splitter into two air ducts passing along the fuselage sides, flanking the cockpit. A conical radome is provided at the upper intake lip/splitter junction, plus a hemispherical radome on the intake splitter; they house the search and tracking antennas of the Almaz-3 radar respectively.

The cockpit is enclosed by a two-piece bubble canopy with a fixed windscreen and an aft-sliding rear portion. The latter parts company with the airframe together with the ejection seat in the event of an ejection, serving as a shield for the pilot. Two fuel tanks are located aft of the cockpit between the inlet ducts. The centre portion of the fuselage incorporates wing attachment fittings.

The *rear fuselage* is a one-piece structure accommodating the engine with its extension jetpipe and afterburner. The rear fuselage incorporates the No.3 fuel tank, four airbrakes and a ventrally located brake parachute container.

Wings: Cantilever mid-wing monoplane with delta wings. Leading-edge sweep 60°, anhedral 2° from roots, incidence 0°, no camber; aspect ratio 2.148, taper 27.7. The mean aerodynamic chord (MAC) is 5,122.7 mm (16 ft 9¹¹⁄₁₆ in).

The wings are one-piece structures utilising a TsAGI S-9S symmetrical airfoil. The wings incorporate the mainwheel wells. The trailing edge is occupied by hydraulically actuated one-piece Fowler flaps and ailerons; maximum aileron deflection is ±15° and maximum flap setting 20°. Each wing has a single pylon with a missile launch rail.

Tail unit: Conventional tail surfaces; sweepback at quarter-chord 55°. The vertical tail comprises a one-piece fin and an inset rudder.

The cantilever horizontal tail consists of slab stabilisers (stabilators) rotating on axles set at 48°30' to the fuselage axis; dihedral 5°, aspect ratio 1.01, travel limits 5° up and 17° down. The stabilators use a TsAGI S-11S-6 symmetrical airfoil.

Landing gear: Hydraulically retractable tricycle type, with single wheel on each unit; the nose unit retracts forward, the main units inward into the wing roots. All three landing gear struts have levered suspension and oleo-pneumatic shock absorbers; the nose unit is equipped with a shimmy damper. The nose unit is castoring; steering on the ground is by differential braking.

The steerable nose unit is equipped with a 570 x 140 mm (22.4 x 5.5 in) K-283 non-braking wheel and a shimmy damper. The main units have 800 x 200 mm (31.5 x 7.87 in) KT-50/2 mainwheels (ko*leso* tor*moznoye*) equipped with disc brakes. The nosewheel well is closed by twin lateral doors, the mainwheel wells by triple doors (one segment is hinged to the front spar, one to the root rib and a third segment attached to the oleo leg). All doors remain open when the gear is down.

Powerplant: One Lyul'ka AL-7F axial-flow afterburning turbojet with a specified thrust of 7,500 kgp (16,530 lbst) at full military power and 10,000 kgp (22,045 lbst) in full afterburner; the actual thrust was considerably lower. A detailed description of this engine is found in the structural description of the Su-9 (see Chapter 7).

Fuel system: Internal fuel is carried in two fuel cells in the forward fuselage and two integral tanks in the wing torsion box (aft of the mainwheel wells) holding a total of 3,130 litres (688.6 Imp gal); the fuel load is 2,570 kg (5,665 lb). There are provisions for carrying two 500-litre (110 Imp gal) drop tanks on pylons under the centre fuselage holding an extra 820 kg (1,810 lb) of fuel.

Armament: Two K-7L or K-6V semi-active radar homing air-to-air missiles carried on pylon-mounted launch rails under the wings. Provisions were made for installing two 30-mm Nudelman/Rikhter NR-30 cannons with 65 rpg in the wing roots but these were never installed.

Avionics and equipment: The avionics suite includes an Almaz-3 fire control radar, an RSIU-4 Doob two-way VHF communications radio (later replaced by a Mindal' radio but then changed back to the previous model), an MRP-48P marker beacon receiver (**mar**kernyy rahdiopree**yom**nik), an ARK-5 automatic direction finder (avtoma**ti**cheskiy **rahdio**kompas), a Sirena-2 radar warning receiver, a GIK-1 gyro-flux gate compass (**ghee**roinduk**tsion**nyy **kom**pas), an AGI-1 fighter-type non-toppling artificial horizon (aviagorizont istrebi**tel**'nyy), a PVU-67 computing sight and an SRZO-2M Kremniy-2M IFF interrogator/transponder.

An artist's impression of the PT-7, 'the horror borne on the wings of the night'. The twin pointed radomes and the raked air intake leading edge gave the aircraft a positively hair-raising look.

PT-7 experimental interceptor

In 1955-56 OKB-51 continued work on adapting the T-3 interceptor for carrying K-7L or K-6V AAMs. The second prototype, known as *dooblyor* (lit. 'understudy'; this was the Soviet term for second prototypes used until the late 1960s), received a separate designation, PT-7, and was designed to be armed with missiles from the start. (PT probably stood for pere**khvaht**chik s treu**gol**'nym k**rylom** – delta-wing interceptor.) The PT-7 was to be equipped with an improved Almaz-7 fire control radar and a PVU-67 computing sight instead of an ASP-5N optical sight (*avtoma**tich**eskiy stre**lkov**yy p**ritsel**). The installation of a new radar required changes to the forward fuselage; the tracking antenna was moved down to the air intake's lower lip and enclosed by a new conical radome, and the effect was rather bizarre, the aircraft looking like a misshapen elephant with two tusks. The air intake leading edge was no longer vertical but angled 16° aft.

The missiles were carried on underwing pylons, one under each wing; no cannons were fitted. In this guise the aircraft was to enter mass production.

Detail design was completed in December 1955, prototype construction commencing at the end of that year. The changes were not limited to the new radar and attendant 'nose job'; the wing design was altered in accordance with TsAGI recommendations. This, together with the late delivery of the engine, caused the PT-7 to be completed much later than intended. At the time the OKB was in the process of testing the T-3 and S-1 prototypes which, as already mentioned, were powered by the AL-7F – the first afterburning version of this engine. Soon, however, OKB-165 proposed an uprated version of this engine which in due course was designated AL-7F-1. The idea was supported, and on 25th August 1956 the Council of Ministers issued a directive requiring Sukhoi's OKB-51 to install the AL-7F-1 on the T-3 and S-1 with a view to ensuring a service ceiling of at least 21,000 m (68,900 ft).

Once again the Sukhoi OKB had to make changes to the manufacturing drawings. The AL-7F-1 had a slightly larger casing diameter and was too big to fit inside the existing rear fuselage, which had to be widened somewhat; this change concerned both the interceptor and the S-1. This meant further delays while MMZ No.51 manufactured a new rear fuselage section, and the PT-7 was not delivered to Zhukovskiy until early June 1957. Eduard V. Yelian was appointed project test pilot, with K. N. Strekalov as engineer in charge of the test programme.

The maiden flight took place at the end of June 1957. The manufacturer's flight tests involved 24 flights to check the aircraft's performance and handling and verify the principal systems; after that, on 23rd September the PT-7 was ferried to the GK NII VVS facility at Vladimirovka AB. Since the T-3's State acceptance trials schedule proved impossible to keep, on 13th September the Council of Ministers issued a new directive setting a new deadline for the commencement of the trials (December 1957). The Sukhoi OKB did its utmost to complete the manufacturer's tests of the K-7L weapons system, including live launches, before the aircraft was handed over to the military. However, these plans were shattered. After making only two flights under this programme the aircraft was grounded for yet another engine change; a replacement engine was not delivered until November, which is why a mere six flights could be made until the end of the year. The work continued in 1958, another 18 flights following by the end of June in which only six missile launches were made; the PT-7 was flown by OKB-51 test pilots L. G. Kobishchan and A. A. Koznov.

PT-8 interceptor (*izdeliye* 27)

Since the T-3 interceptor was expected to enter production in its PT-7 configuration, MAP issued an order to the effect that aircraft factory No.153 named after Valeriy P. Chkalov in Novosibirsk should manufacture three pre-production examples in 1956, delivering them to OKB-51 in order to widen the scope of development work and accelerate the trials. The production version of the T-3/ PT-7 was allocated a new designation, PT-8. At that time plant No.153 was mass-producing the MiG-19S tactical fighter which had the internal designation *izdeliye* 26; since the PT-8 was next in line, it received the product code *izdeliye* 27.

The time schedule proved to be a little too optimistic, considering that the mastering of a completely new type (and an aircraft coming from an OKB with whose products the plant had no prior experience) was involved. Bearing the construction number 0015301 which followed the traditional system in use at Novosibirsk (ie, batch 00, plant No.153, 01st aircraft in the batch; often quoted simply as 0001), the first pre-production aircraft was completed in February 1957. The other two machines (c/ns 0015302 and 0015303) followed in the spring of that year. All three aircraft were dismantled straight away and delivered by rail to OKB-51 in Moscow where they were to be flown. Eventually, however, only one of the three ever flew in as-built configuration, the other two being extensively modified before they entered flight test. The aircraft in question was flown by LII test pilot V. M. Pronyakin and was to participate in that year's Tushino air parade on Aviation Day but the event was cancelled at the last moment.

In April 1958 the Council of Ministers issued a new directive specifying the performance targets for the T-3 interceptor. Importantly, the new SOR envisaged that a new armament system was to be integrated in the course of further work.

T-39 interceptor project

Striving to increase the T-3's service ceiling, in 1958 OKB-51 started work on the T-39 development aircraft; this was intended to explore the possibility of increasing engine thrust by means of water injection into the afterburner (more probably, not pure water but a water/methanol mixture was to be injected). The third pre-production T-3 (c/n 0015303) – ie, the PT8-3 – was selected for conversion as the T-39. The conversion involved replacing the No.3 fuel tank in the rear fuselage with a 700-litre (154 Imp gal) tank for the water injection system; to compensate for this, an extra fuel tank was provided in the fuselage nose.

The aircraft was never tested in this guise because the water injection system programme was transferred to the Central Aero Engine Institute (TsIAM – *Tsentrahl'nyy institoot aviatsionnovo motorostroyeniya*) for further research, using ground test rigs. The unflown T-39 was further converted to become the T-49 development aircraft described later in this book.

T-43 (T43-1) research aircraft

Starting in the mid-1950s, US high-altitude reconnaissance aircraft began intruding ever more frequently into Soviet airspace. This led the Soviet Union to accelerate development of not only surface-to-air missiles but manned interceptor aircraft as well. The Central Committee of the Soviet Union's Communist Party held a session attended both by top-ranking Ministry of Defence officials and by representatives of most defence industry branches. On 25th August 1956 the Council of Ministers let loose with a huge directive requiring all Soviet fighter design bureaux to increase the service ceiling of their new fighters within an extremely brief period. In particular, Pavel O. Sukhoi was ordered to increase the service ceiling of the S-1 fighter and the T-3 interceptor (both of which were then under development) to 21,000 m (68,900 ft). To this end both aircraft were to be powered by the uprated AL-7F-1 engine and the OKB was authorised to delete some systems of secondary importance for the purpose of cutting the interceptor's empty weight.

As noted earlier, the AL-7F-1 had a slightly larger casing diameter, necessitating a redesign of the detachable rear fuselage section which was widened accordingly. A while earlier, acting on the recommendations of TsAGI, the engineers had altered the wing design to incorporate a leading-edge dogtooth; at high angles of attack this dogtooth was to generate a vortex which limited spanwise airflow, delaying tip stall and increasing wing lift. Another change resulted from the early flight test results obtained with the T-3 prototype – aileron area was reduced.

Design work on the altered airframe components of the interceptor was completed in December 1956; the manufacturing drawings for these were issued both to MMZ No.51 (the OKB's prototype construction shop) and the production factory No.153 in Novosibirsk so that appropriate changes could be made to production T-3s (PT-8s). (At that point the decision to launch full-scale production had already been taken and the production plan for 1957 included 30 PT-8s.) As a 'belt and braces policy', MAP had agreed with the customer (the Air Force) that provisions would be made for equipping the initial-production aircraft with the old AL-7F engine and NR-30 cannons in case the envisaged AL-7F-1 engine and the K-7 air-to-air missile were not available on schedule. Additionally, MAP issued a series of orders requiring each of the Soviet 'fighter makers' to perform research and development work with the purpose of increasing the service ceiling. This accounts for the sudden interest in rocket boosters displayed – yet again – by the Soviet aircraft industry in the late 1950s. Specifically, the Sukhoi OKB was instructed to equip the basic T-3 interceptor with two alternative rocket boosters – the U-19 (U = *ooskoritel'* – booster) built around the S3-20 liquid-fuel rocket motor designed by L. L. Sevrook at the Ministry of Defence Industry's OKB-3 and the U-19D based on the RU-013 reusable rocket motor designed by L. S. Dooshkin (hence the D). Water injection into the afterburner as a means of increasing thrust was also considered.

The Sukhoi OKB decided to fit the rocket booster concurrently with the AL-7F-1 turbojet for which the aircraft would act as a testbed. The mixed-power derivative of the T-3 received the manufacturer's designation T-43.

In addition to the powerplant, the T-43 featured a new air intake design. The high speeds specified by the said CofM directive could not be attained with the T-3's fixed-area subsonic air intake having rounded lips, as this design incurred considerable pressure losses. Hence, teaming up with TsAGI, the Sukhoi OKB had been working on a more efficient intake design since 1955. The engineers

Above: The T-5 development aircraft featured a sharp-lipped axisymmetrical air intake with a shock cone as used on the S-1 and T-43. Note the gun blast plates.

This rear view of the T-5 illustrates the new rear fuselage housing two Tumanskiy R11F-300 engines side by side.

decided to use a variable air intake utilising a series of sloping shock waves to provide gradual deceleration of the airflow with minimum pressure losses. However, the need to accommodate the bulky fire control radar was still there.

The most obvious solution was to use an axisymmetrical conical centrebody (shock cone) as used on the S-1 fighter, with a compound shape creating two shock waves – a so-called two-shock cone. Unlike the preliminary development project of the S-1, the circular axisymmetrical air intake featured sharp lips. The centrebody was movable, featuring two main positions; when the aircraft accelerated to Mach 1.35, an automatic control system triggered by an airspeed sensor moved the shock cone to full forward position, retracting it into full aft position at lower speeds. The OKB decided to test this new intake design on the T-43 as well.

Since the T-43 was purely an experimental aircraft, no provisions were made for armament or mission avionics. The avionics bay and the space inside the air intake centrebody were occupied by test equipment and ballast to maintain the correct CG position.

The OKB's experimental shop manufactured three sets of ventral rocket booster housings for the T-43; however, development of the boosters proper was running behind schedule and OKB-51 chose to begin initial flight tests in as-is condition (*sans* booster). Vladimir S. Ilyushin was appointed project test pilot, while M. I. Zooyev was engineer in charge. The T-43's airframe was completed in the late summer of 1957 but remained at the factory for another month pending delivery of a flight-cleared AL-7F-1. At the end of September the still engineless aircraft was trucked to Zhukovskiy. On 1st October OKB-165 finally delivered the engine, which was promptly installed, and on 10th October the T-43 successfully performed its maiden flight.

In its third flight, on 20th October, the T-43 confirmed the promise it held by climbing to 21,500 m (70,540 ft) – higher than the government directive required – on the power of the main engine alone (no booster was fitted). Three days later the aircraft excelled again, clocking a speed of 2,200 km/h (1,366 mph), which was equivalent to Mach 2.06. This success came at exactly the right moment, as the MAP bosses needed something to show the head of state, Nikita S. Khrushchov (who was becoming increasingly scornful in his attitude towards military aviation and biased towards missile systems), that manned combat aircraft were still a force to be reckoned with. The ministry issued a special order commending Vladimir S. Ilyushin for this performance; meanwhile, the Sukhoi OKB was instructed to install a fire control radar in an interceptor

This view of the T-5 shows well how the wing flaps are cut away for integration with the 'hips' of the substantially widened rear fuselage. Note the engine bay cooling air intakes and the forward avionics bay cover ahead of the windscreen.

featuring the new air intake design. Nobody mentions rocket boosters or water injection anymore.

In the course of later trials the T-43 (subsequently redesignated T43-1 because other examples appeared) was retrofitted with an ESUV-1 electrohydraulic air intake control system (e*lek*troghidrav*lich*eskaya *sistema oo*prav*len*iya vozdookhoza*bornikom*) designed to prevent engine surge; the system provided continuous control of the shock cone throughout the flight envelope. In subsonic flight the shock cone was still in the full aft position but then moved forward gradually as the speed increased past Mach 1, ensuring the optimum position of the shock waves. This system later found use on other aircraft in the T-3 series featuring axisymmetrical air intakes.

T-5 experimental interceptor (*izdeliye* 81-1)

To meet an Air Force requirement the Sukhoi OKB developed a much-modified version of the T-3 designated T-5 or *izdeliye* 81-1. The aircraft was powered by two R11F-300 axial-flow afterburning turbojets developed by Sergey K. Tumanskiy's OKB-300, with a take-off thrust of 4,200 kgp (9,260 lbst) dry and 6,120 kgp (13,490 lbst) reheat, and was effectively a propulsion testbed designed to verify this twin-engine powerplant.

Being lighter and more compact than the AL-7F-1, the Tumanskiy engines were installed side by side in a completely redesigned and much wider rear fuselage; the result was a pronounced 'waist' at the wing trailing edge. The fuselage break point was moved aft from frame 28 to frame 34. The section between frames 28 and 34 was a sort of adapter between the existing forward fuselage structure and the new rear fuselage, incorporating a bifurcated inlet duct for the two engines; it also housed an enlarged No.3 fuselage tank increasing the total internal fuel capacity from the T-3's 3,130 litres (688.6 Imp gal) to 3,480 litres (765.6 Imp gal). (In reality, however, not more than 3,330 litres (732.6 Imp gal) was filled during trials so as not to exceed the maximum take-off weight, the missing 150 litres (33 Imp gal) being distributed between the wing tanks.)

Other structural changes included a new fuselage nose with an axisymmetrical air intake identical to the one tested on the T43-1. The inboard ends of the flaps were slightly cropped to cater for the increased width of the fuselage. Unlike the AL-7, which was started by a jet fuel starter (a small gas turbine engine), the smaller and lighter R11-300 had electric starting; hence the single 12-kilowatt GS-12T generator was replaced by two 9-kilowatt GSR-ST-9000A starter-generators. Finally, the BU-30 and BU-34 hydraulic control surface actuators were replaced with identical BU-49 actuators in all three control circuits, and the control cables in the rear fuselage gave place to push-pull rods. The rest of the equipment remained unaltered.

The T-5 was converted from the T-3 prototype (c/n 01) at MMZ No.51. Due to the extent of the changes the conversion job took eight months (from October 1957 to June 1958). In early July the aircraft was delivered to the flight test facility in Zhukovskiy to commence ground checks, with M. I. Zooyev as engineer in charge of the tests. On 18th July the T-5 made its first flight at the hands of Vladimir S. Ilyushin. The manufacturer's flight test programme was completed in full, involving 26 flights. The tests showed that, with the engines in afterburner mode, the aircraft was overpowered and, in spite of the decidedly higher drag (primarily due to the wider rear fuselage), could reach much higher speeds than the T-3; the engines' structural strength was the limiting factor, the R11F-300 being designed to withstand speeds below Mach 2.

On the down side, the engines' automatic fuel control units still had a few bugs to be eliminated; as a result, the engines ran unstably, the afterburners shutting down frequently of their own accord during climb. Engine starting proved problematic as well, since both engines used a common air intake and, figuratively speaking, were short of breath. (It was much the same story with the MiG-19 where the two engines breathed through a single air intake divided by a splitter into individual inlet ducts; the downwind engine had to be started first, otherwise the other engine would literally take all the air away from it!) Besides, the T-5's longitudinal stability proved to be unacceptably low (to be precise, the aircraft had virtually zero stability due to the CG being positioned well aft). In May 1959 the aircraft's chief project engineer Yevgeniy Fel'sner called a halt to the test programme.

PT-95 testbed

The second pre-production T-3 (c/n 0015302) – ie, the second PT-8 (the PT8-2) – never flew as such but was immediately converted into a propulsion testbed designated PT-95. Receiving a specially instrumented engine with test equipment sensors and a new forward fuselage patterned on the T43-1, the aircraft was delivered to its new owner, LII, in 1958. In 1958-59 the PT-95 served as a testbed for the new AL-7F-1, helping the engine's teething troubles to be overcome, and was used to explore the interaction between the air intake and the engine. The programme was performed by LII test pilot V. P. Vasin; on one occasion he had to make a dead-stick landing in the PT-95.

T-3-51 aerial intercept weapons system

T43-2, T43-3, T43-4, T43-5, T43-6 and T43-11 interceptor fighter prototypes

After the success of the T43-1 development aircraft the Sukhoi OKB concentrated on two main areas of work – attempts to achieve a satisfactory radar installation within the small shock cone used on the T43-1 and studies on accommodating the Almaz radar in a new and larger shock cone. The latter effort culminated in a new project designated T-47 which is described in Chapter 2.

In the mid-1950s the NII-17 research institute was the Soviet Union's sole maker of aircraft radars. Only two principal airborne intercept radar types developed by NII-17 – the RP-1 *Izumrood-1* (Emerald-1) twin-antenna radar (plus its refined version designated RP-5 Izumrood-5) and the RP-6 **Sokol** (Falcon) single-antenna radar – had entered production by 1957; the Izumrood series radars were fitted to the MiG-17P, MiG-17PF, MiG-17PFU, MiG-19P and MiG-19PM, while the more powerful Sokol equipped the Yakovlev Yak-25M, Yak-25K and Yak-27 interceptors. (RP = **rah**dio**pri**tsel – 'radio sight', the Soviet term for fire control radars.) However, neither of these radars was suitable for an advanced supersonic interceptor; the RP-1/RP-5 offered inadequate performance, while the RP-6 was too bulky to fit inside the shock cone of an adjustable supersonic air intake. True, NII-17 had a couple of aces up its sleeve – the new and advanced Uragan (Hurricane, pronounced *ooragahn*) and *Pantera* (Panther) fire control radars, but their development was making painfully slow progress.

It was then that a competitor popped up like a jack-in-the-box; OKB-1, a division of the Ministry of Defence Industry (MOP – *Ministerstvo obo***ron***noy pro***mysh***lennosti*), suddenly emerged as a new airborne radar design house. Putting the know-how gained in the development of air-to-surface missile guidance systems to good use, an OKB-1 design team headed by project chief A. A. Kolosov had quietly developed the relatively compact TsD-30 fire control radar. The TsD-30 was optimised for guiding the K-5 semi-active radar homing missile created by Pavel D. Grooshin's OKB-2 – and, importantly, it was small enough to fit inside the T-43's movable shock cone without any trouble. Another major point in favour of this radar was that the K-5 AAM (known in service as the RS-1-U, ie, ra***ket***nyy sna***ryad***, tip ***odin***, ooprav***lya***yemyy – missile, Mark 1, guided) was the only air-to-air missile included into the Soviet Air Force's inventory by 1957. In October that year an improved version designated K-5M successfully passed checkout trials on a MiG-19PM interceptor. The proper conclusions were drawn in high places, and on 28th November the Council of Ministers issued a directive requiring OKB-51 to equip the AL-7F-1 powered T-3 interceptor with the TsD-30 radar and arm it with K-5M missiles.

The follow-up CofM directive that appeared on 16th April 1958, in effect, ordered the development of the T-3's production version, formulating the requirements for what the aircraft eventually became. The aircraft was no longer regarded merely as an interceptor; it was to form part of an aerial intercept weapons system comprising the aircraft proper and the **Voz***dookh*-1 (Air-1) ground controlled intercept (GCI) system.

The T43-15 prototype with a full load of external stores – four K-5MS missiles and two 600-litre (132 Imp gal) drop tanks. Note the two cine cameras under the nose to record missile launches.

Air Marshal Yevgeniy Yakovlevich Savitskiy, Commander of the Air Defence Force's fighter arm, participated in the tests of Sukhoi delta-wing jets.

Test pilot Vladimir N. Makhalin, Hero of the Soviet Union. He performed the maiden flight of the T-3, the progenitor of the Sukhoi deltas.

The directive envisaged two versions of the interceptor; one was equipped with the TsD-30 radar and armed with four K-5M AAMs, while the other featured an Oryol (Eagle) radar and two K-8M AAMs. The weapons systems built around these two versions of the aircraft were designated T-3-51 and T-8M respectively (the latter is described in the next chapter). OKB-51, the main organisation responsible for the development of both weapons systems, was also tasked with creating a combat-capable trainer version of the aircraft. All further work on the K-7L and K-6V missiles was terminated.

The directive set a fairly tight development schedule – the aircraft were to be submitted for State acceptance trials in the third quarter of 1958. For the first time in Soviet practice, due to the high priority allocated to the programme the OKB was required to submit no fewer than six flying prototypes instead of the usual two.

(It may well be said now that this directive ultimately allowed the two weapons systems to be progressively put into production and fielded, the T-3-51 system first. In practice, however, development and tests of the two versions in 1958-60 proceeded in parallel.)

OKB-51 allocated the existing T-43 product code to the T-3-51 weapons system. In addition to the existing T-43 development aircraft (which now became the T43-1), a further five initial-production T-3s (PT-8s) were to be converted to the new standard. In order to speed up the process it was decided to convert two of the aircraft in Moscow (at MMZ No.51), while the others would be modified *in situ* at the Novosibirsk aircraft factory. The interceptors were to be equipped with improved Sukhoi KS-2 ejection seats; the new model, which had successfully completed manufacturer's tests in March 1958, featured more effective arm restraints, allowing ejection to take place at indicated airspeeds up to 1,000 km/h (620 mph) instead of 850 km/h (528 mph) for the KS-1. Also, the aircraft were to be fitted with the receiver of the *Lazoor'* (Prussian Blue) data link system forming part of the Vozdookh-1 GCI system.

Meanwhile, OKB-2 adapted the K-5M (RS-2-U) missile which had been tested and perfected on the MiG-17PFU to the T-43; the resulting version was designated K-5MS, the S standing for Sukhoi. Derived from the K-5 (RS-1-U), the weapon was a very basic beam-riding AAM – ie, it followed the beam of the interceptor's fire control radar. Having detected a target on the radar display with the radar operating in search mode, the pilot got the target into the 'crosshairs' in the centre of the display by controlling the aircraft, selected target lock-on mode and, after closing in to the required range, fired the missiles. After that, the pilot was to keep the target in the 'crosshairs' until the missile scored a 'kill'; the missile was controlled by a guidance system monitoring the position of the radar's directional pattern axis.

The K-5MS used a canard layout with nose-mounted control surfaces; it featured two nozzles flanking the rear part of the body, the aft extremity of the latter being occupied by the guidance system's receiver antenna in a dielectric fairing. The missile had a 13-kg (28.5-lb) high-explosive/fragmentation warhead and a proximity fuse. For carrying the missiles the aircraft featured four permanently installed pylons with APU-20 launch rails (*aviatsionnaya pooskovaya oostanovka* – aircraft-mounted launcher). Together with the TsD-30 radar and the launch control equipment the K-5MS missiles formed the K-51 (S-2-US) armament system.

The prototypes converted in Moscow received the designations T43-2 and T43-6, the three examples modified in Novosibirsk being designated T43-3 through T43-5. The T43-2 had started life as the third production PT-8 (c/n 0115303) which was delivered to MMZ No.51 in early February 1958. By the end of May that year OKB test pilot V. N. Il'yin had made the first three flights of the manufacturer's test programme in this aircraft. The first Novosibirsk-built example was converted almost concurrently from the fifth production PT-8 (c/n 0115305), LII test pilot V. M. Pronyakin arriving from Moscow to make the first flight in this aircraft; factory test pilot T. T. Ly-senko also made a familiarisation flight in the T43-3 in the course of pre-delivery tests. In June Pronyakin ferried the aircraft to Zhukovskiy. Concerned about the development schedule being maintained, the MAP top brass kept a close watch on the programme, harassing the OKB; between May and August 1958 alone the T-43 programme came under close scrutiny four times!

A major problem which had been plaguing the T-3 family was eliminated in the summer of 1958: the aerodynamicists at OKB-51 suggested installing auxiliary blow-in doors aft of the air intake to admit additional air and prevent engine surge. The feature was tested on the T43-1 and proved so successful that MAP immediately prescribed such blow-in doors to be incorporated on the Su-7 fighter-bomber and the MiG-21F fighter, both of which had just entered production.

The T43-6 joined the test programme in July 1958, followed by the T43-4 and T43-5 in August; thus all six aircraft envisaged by the CofM directive were available and flying at the end of the manufacturer's test programme. On 30th August 1958 Pyotr V. Dement'yev, Chairman of the State Committee for Aviation Hardware (GKAT – *Gosoodarstvennyy komitet po aviatsionnoy tekhnike*) wrote an official letter to the Soviet Air Force Commander-in-Chief Air Marshal Konstantin A. Vershinin, submitting the T-43 for State acceptance trials. (Note: In December 1957 MAP lost its ministerial status together with several other ministries and was 'demoted' to the State Committee for Aviation Hardware because of the Soviet leader Nikita S. Khrushchov's disdainful attitude. In 1965, when Khrushchov was unseated and replaced by Leonid I. Brezhnev, GKAT regained its original name and 'rank'.)

A State commission chaired by Deputy C-in-C Col. Gen. Fyodor A. Agal'tsov was formed for conducting the trials. However, these could not be started for another three months because the Sukhoi and Lyul'ka

Sukhoi OKB test pilot A. A. Koznov.

Sukhoi OKB test pilot A. S. Komarov.

Sukhoi OKB test pilot Yevgeniy S. Solov'yov.

GK NII VVS test pilot Stepan Anastasovich Mikoyan.

Sukhoi OKB chief test pilot Vladimir Sergeyevich Ilyushin.

GK NII VVS test pilot Gheorgiy Timofeyevich Beregovoy. He went on to become an astronaut.

Sukhoi OKB test pilot L. G. Kobishchan. He was killed on 20th July 1959 in the crash of the T43-6.

Sukhoi OKB test pilot Eduard Vaganovich Yelian.

GK NII VVS test pilot Nikolay I. Korovushkin.

design bureaux were forced to eliminate the numerous defects unearthed in the course of acceptance.

Stage A of the T-3-51 weapons system's State acceptance trials began on 3rd December 1958, continuing until May 1959; this was the so-called General Designer's stage, ie, the flights were performed by OKB personnel. Stage B held jointly by the OKB and the Air Force lasted from June 1959 to April 1960. The T-43 prototypes were flown by OKB-51 test pilots Vladimir S. Ilyushin, A. A. Koznov, L. G. Kobishchan, Yevgeniy S. Solov'yov and N. M. Krylov, as well as GK NII VVS pilots Gheorgiy T. Beregovoy, Nikolay I. Korovushkin, L. N. Fadeyev, Boris M. Andrianov, V. G. Plyushkin and Stepan A. Mikoyan (the Mikoyan OKB founder's nephew). The trials proceeded with a fair share of problems due to the newness of the aircraft as a whole and the inevitable teething troubles of its systems, including the powerplant and the air intake control system, which surfaced at this stage. The engine often surged when throttled back at speeds equivalent to Mach 1.8 and higher, or at speeds above Mach 1.5 at altitudes in excess of 15,000 m (49,210 ft). To cure the problem all six prototypes had the intake centrebody travel increased from 21.5 to 23 cm (from 8^{15}/$_{32}$ in to 9 in) and the ESUV-1 electro-hydraulic air intake control system installed in January 1959 after these measures had been tried successfully on the T43-1. No more cases of engine surge were recorded during the State acceptance trials.

At a conference held in February 1959 to check up on progress, V. P. Belodedenko, who was assigned to the T-43 programme as GK NII VVS's project engineer, said that 'the system is very good. We need it desperately. We believe in this system and we believe it holds great promise and can be brought up to operational level easily, but delays in this process can discredit everything.' Even before the T-43 entered the decisive phase of the State acceptance trials, the engine makers at OKB-165 promised to increase the AL-7F-1's maximum speed to 8,500 rpm (until then the engine had been incapable of doing more than 8,350 rpm) and install the upgraded engine on production aircraft from the third quarter of 1959 onwards. Speaking at the same conference, test pilot Vladimir S. Ilyushin assured the Air Force representatives in attendance that in a few days the T43-2 equipped with auxiliary blow-in doors would demonstrate stable engine operation when intercepting a simulated target at 20,000 m (65,620 ft), with no tendency to flame out when four K-5M AAMs were fired.

On 20th July 1959 the T43-6 crashed during a post-modification checkout flight from Chkalovskaya AB east of Moscow, killing test pilot L. G. Kobishchan. When the pilot stopped responding to the tower, a search and rescue operation was mounted. Presently the SAR team found a deep crater on the outskirts of Serkovo village 4 km (2.5 miles) from the airbase; the aircraft had hit the ground in a nearly vertical dive with tremendous force, burying itself deep in the soft soil. Extracting the wreckage proved impossible because the boggy ground was giving way and the crater kept filling with ground water; only a few fragments were recovered. The cause of the crash was thus never ascertained; however, eyewitnesses reported that the aircraft had started disintegrating in mid-air, which may indicate that the pilot had passed out for some reason and the uncontrollable aircraft had entered a dive in full afterburner, exceeding its never-exceed speed (V_{NE}). A monument was later erected to mark the T43-6's and Kobishchan's place of eternal rest.

To make up for the loss of the T43-6 the Air Force transferred another production T-8 to the Sukhoi OKB; upon conversion this aircraft was redesignated T43-11, joining the State acceptance trials programme in August 1959. Stage B was mostly concerned with determining the aerial intercept weapons system's efficiency and performance. Since no high-altiude targets were available, live missile launches were made at medium altitudes only – mostly against M-15 (MiG-15M) radio-controlled target drones. The OKB wasted no time making the necessary changes to the aircraft when deficiencies were discovered. Thus the DC battery was placed in a pressurised container to ensure stable performance at high altitudes, and the radar display was provided with a rubber sunblind to facilitate sighting.

The State acceptance trials, which included a total of 407 flights, were concluded on 9th April 1960 when the State commission signed the final protocol. The commission pointed out that virtually all of the performance targets set in the Council of Ministers directive had been met. In particular, the aircraft was capable of destroying targets flying at speeds of 800-1,600 km/h (495-990 mph) and altitudes of 5,000-20,000 m (16,400-65,620 ft) with a 'kill' probability of 70-90%; the maximum radius of action (interception range) was 430 km (267 miles) instead of the specified 400 km (248 miles).

Su-9 production interceptor (T-43, *izdeliye* 27, *izdeliye* 34; *izdeliye* 10)

In the autumn of 1960 the T-43 successfully underwent evaluation at the Aircraft Test Centre of the Air Defence Force's fighter arm the fighter arm of the Air Defence Force (IA PVO – *Istrebitel'naya aviahtsiya Protivovozdooshnoy oborony*) in Krasnovodsk, Turkmenia, whereupon on 15th October 1960 the Council of Ministers issued a directive clearing the T-3-51 aerial intercept weapons system for service entry. All of the system's components received new names in so doing; the T-43 interceptor armed with four K-5MS AAMs received the service designation Su-9, the TsD-30 radar became the RP-9U, the missile itself was redesignated RS-2-US and the system as a whole was renamed Su-9-51.

Predictably, series production of the Su-9 was organised at the Novosibirsk aircraft factory No.153 in accordance with a Council of Ministers directive issued back in September 1958 (ie, immediately after the completion of the PT-8 initial production batch). Interestingly, initially the production Su-9 retained the factory code of the original T-3 (*izdeliye* 27) which was later changed to *izdeliye* 34 when a forward fuselage integral fuel tank and other structural changes were made; nevertheless, in paperwork sent to the factory from the Sukhoi OKB in Moscow the aircraft was still referred to as the T-43 – or even as the T-3 (!).

In 1958 plant No.153 built 40 T-3 series aircraft in various configurations, including the T-43; some of them were used for various test and development work. The first aircraft manufactured to full Su-9 standard – the so-called *etalon* (standard-setter) – was c/n 0215310. (Curiously, even though the sometimes quoted full c/n contained the factory number, 153, on the actual aircraft it was stencilled simply as a four-digit number with a dash separating the batch number from the number of the aircraft within the batch – in this case, 02-10.)

Novosibirsk factory test pilots T. T. Lysenko, B. Z. Popkov, V. V. Proveshchayev and others commenced pre-delivery tests of production Su-9s in the autumn of 1958. Since the runway at Novosibirsk-Yel'tsovka was more of a 'ruinway' (it was in extremely poor condition and in need of resurfacing), the fighters were flown to Novosibirsk-Tolmachovo airport in their first flight to undergo the rest of the pre-delivery tests there.

The factory test pilots' job was no easier than that of an OKB test pilot, and losses occurred, too. Proveshchayev was killed in an accident on 20th November 1958 when one of the intake centrebody's attachment points failed, the centrebody tilting upwards and cutting off the air supply to the engine. Five accidents and incidents occurred in the course of pre-delivery tests in 1959. Two of them were caused by engine failures; luckily, on both occasions the pilots (V. A. Bogdanov and Yuriy N. Kharchenko) managed a safe landing at the airfield of origin. That same year V. P. Krooglov, a test pilot of the Air Force's quality control and acceptance team, attempted an emergency landing when the port main landing gear unit refused to extend. On 19th November 1959 factory test pilot

A. D. Dvoryanchikov attempted a dead-stick landing after an engine failure but crashed, losing his life. Another brand-new Su-9 was lost during pre-delivery tests in 1960, test pilot V. T. Vylomov ejecting safely.

Su-9 production in Novosibirsk continued until 1962. Additionally, the interceptor was built in quantity at MMZ No.30 '*Znamya Trooda*' (*Moskovskiy mashinostroitel'nyy zavod* – Moscow Machinery Plant No.30 'Banner of Labour'), a production factory located at the east end of Moscow-Khodynka airfield, only a short way from the Sukhoi OKB's premises. This plant received a full set of manufacturing documents and several shipsets of airframe/ systems components from Novosibirsk in 1959, assembling the first two pre-production aircraft in mid-year; full-scale production commenced the following year, continuing until 1961. The Su-9's in-house product code at MMZ No.30 was *izdeliye* 10.

Between them the two production plants completed just over 1,000 Su-9s. The Novosibirsk factory completed a total of 888 T-3 series aircraft in various guises in 1957-62, while Moscow production totalled 126 single-seaters and 50 Su-9U dual-control trainers described later in this chapter.

The new interceptor was publicly unveiled on 9th July 1961 when a formation of Su-9s took part in the traditional Aviation Day flypast at Moscow-Tushino. The aircraft were flown by instructor pilots from the IA PVO's 148th TsBP i PLS (*Tsentr boyevoy podgotovki i pereoochivaniya lyotnovo sostahva* – Combat & Conversion Training Centre) located at Savostleyka AB near Gor'kiy (now renamed back to Nizhniy Novgorod). Immediately afterwards the Western press published the first pictures of the Su-9; again, intelligence experts unerringly identified it as a Sukhoi aircraft and the Su-9 received the NATO code-name *Fishpot-B*.

In early 1961 LII held a full-scale spinning trials programme with the Su-9. The aircraft was flown by test pilots Sergey N. Anokhin and A. A. Shcherbakov; the latter specialised in spinning trials and had considerable expertise in this field. Spins were initiated at 12,000 m (39,370 ft); mindful of the possibility of engine surge at the moment of spin entry, the pilots usually shut the engine down in advance, restarting it after spin recovery. The aircraft would lose between 2,500 and 3,500 m (8,200-11,480 ft) of altitude between spin entry and recovery. According to the pilots' reports, the Su-9 could enter a spin only due to a grave piloting error – or if the spin was intentional. The spin itself was unstable, the aircraft falling like a maple leaf, with angles of attack (AOAs) around 45-50° and angle speeds of about 1.7 radian per second.

GK NII VVS pilots joined in the spinning trials a while later. On 30th March 1961, when Air Force test pilot L. N. Fadeyev was flying the aircraft, the engine surged and refused to relight after spin recovery. Fadeyev attempted a forced landing on LII's unpaved runway (probably purpose-built for rough-field tests, as Zhukovskiy surely had a concrete runway by then!) but the aircraft turned turtle after hitting a pothole during the landing run, the pilot sustaining serious spinal injuries.

Despite persistent efforts by the OKB and the manufacturing plants to improve reliability, incidents and accidents (including fatal ones) involving production Su-9s were all too frequent. According to the Air Force's attrition statistics, in 1961 alone there were 34 accidents of varying seriousness with the type, including 18 caused by design flaws and manufacturing defects. The mean time between accidents was 677 hours overall and 1,278 hours for those caused by manufacturing defects. By comparison, the Su-7 enjoyed a much better safety record, with only five accidents in 1961 (albeit four out of five were again due to materiel defects); the mean time between accidents was 1,561 and 1,952 hours respectively.

There were many causes for the high accident rate. For instance, no fewer than six crashes were caused by disintegration of the engine's accessory gearbox drive shaft bearings; 30 engines had to be removed prematurely when metal shavings were detected in the lubrication system. By 1961, 35 cases of compressor blade failures on AL-7F and AL-7F-1 engines due to foreign object damage were recorded since the Su-7 fighter, the Su-7B fighter-bomber and the Su-9 had entered service; turbine blade cracking and failures were common, too.

The blame did not rest solely with the engine makers; the poorly maintained runways at airbases were a major contributing factor. The new aircraft operated from old concrete runways which were not designed to withstand the loads generated by the aircraft; the Sukhoi jets had a runway loading of about 12 kg/cm² (171 lb/sq.in). The augmented loads, coupled with the jet exhaust, gradually eroded the runway surface and the top layer disintegrated, whereupon small fragments of concrete were ingested by the engine. The resulting dents on the compressor blades acted as stress concentrators and caused the blades to break – often with tragic consequences. In fact, back in 1959, when serious design flaws had been discovered in the AL-7F-1, someone had proposed throwing this engine out and substituting it with the VK-13 afterburning turbojet developed by Vladimir Ya. Klimov, since this engine did not require major structural changes. A different solution was found eventually: the paved runways were reinforced and it became standard operational procedure to sweep them before a flying session, using powerful 'vacuum cleaners' mounted on lorries to remove any debris.

Other widespread bugs afflicting the Sukhoi jets included mainwheel tyre explosions, deformation of the wing structure and the variable nozzle petals. Worst of all, these defects undermined the image of the aircraft as a whole; the 'head office' (ie, OKB-51), not

A typical early-production Su-9 with no auxiliary blow-in intake doors. The object aft of the nose gear unit is probably an open access panel.

the suppliers of this or that component, got the blame.

Alarmed by this situation, the military started 'knocking on every door', demanding that something be done about it. In keeping with orders coming from the Communist Party Central Committee, in February 1961 a special commission – one might say an investigative commission – started working in IA PVO units, studying the operational peculiarities of the new Sukhoi jet. The commission established that '...*in the course of wide-scale Su-9 operations, new defects and shortcomings adversely affecting [...] the aircraft's reliability and flight safety are still revealed. In 1960-61 there have been three fatal accidents and three non-fatal accidents leading to total hull losses, all caused by engine or engine accessory failures; one fatal accident and three non-fatal accidents/total hull losses caused by failures of off-the-shelf components (AGD-1 and AGI-1 artificial horizons etc.) and one fatal accident and two non-fatal accidents/total hull losses caused by piloting errors or improper maintenance. [...] The flight and ground crews operating the Su-9, which is, in effect, a 'manned missile' (sic – Auth.), have the same pay rates as [...] personnel operating subsonic aircraft, which causes justified complaints and does not promote the service introduction of new aircraft.*'

'The commission hath spoken', and Su-9 operations came to an almost complete standstill. The aircraft industry showed no reaction at first, but the accidents continued, and the military finally ran out of patience. In July 1961 the acting PVO Commander-in-Chief General Gheorgiy V. Zimin reported to Council of Ministers Vice-Chairman Dmitriy F. Ustinov:

'*Only six to eight pilots in each regiment – the best pilots in the unit – keep flying, using the best aircraft available, but operations are still accompanied by a high accident rate due to hardware failures. On 17th June this year a non-fatal accident occurred [...] due to disintegration of a bearing in the engine accessory gearbox, with an ensuing in-flight shutdown. On 11th July a non-fatal accident occurred [...] due to compressor blade failure and engine surge on take-off. On 13th July a fatal accident occurred. Immediately after take-off the pilot reported that smoke was pouring into the cockpit, the afterburner would not shut down and the aircraft had started rolling uncontrollably; the pilot could not eject due to insufficient altitude and was killed. The 19th April 1961 ruling of the Military Industry Commission* (a standing committee on defence industry matters established by the Presidium of the Soviet Council of Ministers – *Auth.*) – *specifically, the item concerning the upgrading of [...] engines installed in operational Su-9s, – is disregarded by the factories.*'

Twenty-one non-fatal accidents and one fatal crash occurred in 1962, followed by 23 more accidents in 1963. The situation appears strange; considering that the Su-9 and Su-7 shared the same engine and many equipment items, theoretically their operational reliability should be on a comparable level but this was not the case. Perhaps the answer lies in the fact that the Su-9, as already mentioned, was built in Novosibirsk and Moscow, whereas the Su-7 and its derivatives were manufactured solely by aircraft factory No.126 in Komsomol'sk-on-Amur in the Soviet Far East.

Three years later the situation had improved, as the main bugs had been ironed out in the course of production; in 1965 the mean time between accidents rose to a palatable 14,274 hours. Despite the fact that the most critical design changes aimed at improving reliability were promptly introduced on the production lines, a few early-production aircraft did find their way to service units in as-was condition. As early as August 1961, however, they were either withdrawn from use or upgraded *in situ* in accordance with OKB bulletins.

The principal changes introduced into the design of the Su-9 in the course of production are listed below:

• Four auxiliary blow-in doors were added on the sides of the nose a short way aft of the air intake leading edge. This change was introduced on the Novosibirsk production line at an early stage; the aircraft built previously were either updated accordingly or struck off charge.

• All surviving Su-9s were retrofitted with SARPP-12 flight data recorders (*sistema avtoma***tich***eskoy reghi***strah***tsii*̈ *parahmetrov polyo***ta** – automatic flight parameter recording system).

• The radar was modified to improve target tracking reliability.

• The PVD-5 pitot (*pree***yom***nik voz***doosh***novo dav***le***niya*) was replaced on all aircraft with the standardised PVD-18 ensuring more accurate altitude determination.

• An AP-28Zh-1B autopilot and red cockpit lighting were introduced.

• The AL-7F-1 engine was replaced by the improved AL-7F1-100 offering a marginally higher thrust and a longer service life.

• The K-283 non-braking nosewheel was replaced with a KT-51 brake-equipped wheel (*koleso tormoznoye*) and the KT-50U mainwheels by KT-89 mainwheels, all having identical dimensions to the earlier models.

• From Su-9 c/n 0715302 onwards the No.1 bag-type fuel tank (fuel cell) in the forward fuselage was replaced by an integral tank patterned on that of the T43-12 development aircraft. Concurrently the sloping fuselage frames adjacent to this tank (which were originally intended to serve as guides for the cannons' ammunition belts) gave place to ordinary ones set at right angles to the fuselage waterline and the former cannon bays in the wing roots were transformed into integral fuel tanks. As a result, total internal fuel capacity was increased from 3,060 litres (673.2 Imp gal) to 3,780 litres (831.6 Imp gal).

• The heat-resistant cannon blast plates ahead of the wing roots were deleted.

• The KS-1 ejection seat was replaced first with the KS-2 and then, on the final batches, with the KS-2A. Later all Su-9s were retrofitted with more refined KS-3 ejection seats.

The interceptor's armament received a lot of attention. The original intention was to complement the RS-2US SARH missiles with the then-new K-13 IR-homing AAMs (the K-13 was a reverse-engineered version of the US AIM-9 Sidewinder, several examples of which had been captured in Vietnam and delivered to the Soviet Union by the Vietnamese). This weapon was tested both on Mikoyan fighters and on Sukhoi aircraft (see below). Eventually, however, operational aircraft were armed with an IR-homing version of the RS-2US featuring an identical IR seeker head. Originally known as the K-55, this missile was redesignated R-55 in service and the modified radar compatible with the R-55 was designated RP-9UK.

In 1966-67 two production Su-9s underwent a two-stage trials programme with a view to using the Su-9 as a strike aircraft armed with bombs (!). The ordnance load recommended for service in strike configuration comprised a mix of two 250-kg (550-lb) FAB-250 high-explosive bombs (*foogahs***naya** *avia***bomba***) and two RS-2US AAMs. The UPK-23-250 pod (*oonifit***see***rovannyy push***ech***nyy kon***teyner** – standardised cannon pod) containing a 23-mm (.90 calibre) Gryazev/Shipoonov GSh-23 twin-barrel cannon with 250 rounds was tested on the Su-9 in the late 1960s and early 1970s. The pod was carried on one of the fuselage hardpoints normally used for drop tanks; no drop tanks were carried in this case. The trials programme performed by GK NII VVS test pilots S. A. Lavrent'yev, V. V. Migoonov and V. K. Ryabiy included gunnery attacks against paradropped targets, Lavochkin La-17 target drones and ground targets. Yet, even though the cannon pod was undoubtedly an asset, the impossibility to carry drop tanks impaired the range unacceptably and the UPK-23-250 did not gain wide wide use on the Su-9.

In 1963 two Su-9s were involved in an effort to explore the possibility of operating the type from unpaved tactical strips. Building on the results of this test programme, IA PVO units received appropriate recommendations concerning such operations.

Until the late 1960s, when the Soviet Air Force started taking delivery of the new

This Su-9 coded '14 Red' and lacking missile pylons was one of the test and development aircraft operated by LII; the exact nature of the test work is unknown.

MiG-25P interceptor, the Su-9 was the Soviet Union's fastest and highest-flying combat aircraft.

From the start of the Su-9's series production a number of aircraft were used in assorted research and development programmes held by MAP, the Air Force and other agencies. Refining the T-3-51 weapons system alone involved about 20 T-43 development aircraft. Other aircraft were used as 'dogships' for verifying structural changes, systems and avionics testbeds, weapons testbeds and control configured vehicles (CCVs). Some of these are listed below.

T43-3, T43-4, T43-5 and T43-8 development aircraft

Upon completion of the T-3-51 weapons system's State acceptance trials the T43-3, T43-4, T43-5 and T43-8 served as weapons testbeds for IR-homing air-to-air missiles. Also, the T43-4 was used to test a so-called pre-emptive starting system which precluded engine flameout caused by missile exhaust gas ingestion when RS-2-US missiles were fired.

T43-7 and T43-10 development aircraft

In 1960-61 the T43-7 and T43-10 served as avionics testbeds for the AP-28Zh-1 autopilot. The latter aircraft manufactured in 1960 was later damaged beyond repair in an accident, so a further aircraft (a production Su-9) was fitted with the autopilot in order to continue the trials. Like the pre-emptive starting system mentioned above, the autopilot later became a standard fit on production Su-9s.

T43-12 development aircraft

To meet the Air Force's demand that the interception range be increased, another Su-9 was modified by replacing the No.1 fuselage fuel cell with an integral tank and incorporating additional wing tanks. Designated T43-12, the modified aircraft first flew in January 1960. As already mentioned, this modification increased internal fuel capacity from 3,060 to 3,780 litres and was incorporated on late-production Su-9s.

T43-5 and T43-12 weapons testbeds

In accordance with an Air Force requirement the Sukhoi OKB modified the T43-5 and T43-12 for testing a new armament arrangement: the wingtips were cropped for the purpose of installing tip-mounted launch rails for K-13 AAMs. (At the same time the Mikoyan OKB conducted a similar test programme with the Ye-6-2 prototype, a precursor of the MiG-21F.) The new arrangement was not introduced on production aircraft, as it was deemed easier to equip the K-5MS (RS-2-US) missile with an IR seeker head.

In 1962 the T43-5 and T43-12 began a new round of trials, now armed with the K-55 IR-homing derivative of the K-5MS. Due to development problems with the missile the trials dragged on for a long time and were not completed until 1967. Eventually, as already mentioned, the R-55 missile was recommended for service along with the RS-2-US.

T43-2 and T43-15 avionics testbeds

In 1960-61 the T43-2 and T43-15 (the latter was converted from a production Su-9, c/n 1115310) served as testbeds for the modified TsD-30TP fire control radar. This version did not enter production and operational Su-9s were still equipped with the standard TsD-30.

T43-17 avionics testbed

Another aircraft bearing the in-house designation T43-17 was used to test a new avionics suite subsequently fitted to the Su-11 interceptor. No details are known.

Su-9 cockpit lighting testbed

In 1961 a Moscow-built Su-9 (c/n 100000308 – ie, *izdeliye* 10, year of manufacture 1960, MMZ No.30 (the first digit is omitted for security reasons to confuse would-be spies), Batch 003, 08th aircraft in the batch) was used for testing a new red cockpit lighting system making the aircraft less observable to enemy aircraft at night.

Su-9 ram-air turbine testbeds

In 1963-64 two Su-9s were fitted experimentally with an ATG-2 ram-air turbine (RAT) providing electric power in an emergency. The RAT supplanted the ventral brake parachute bay and the parachute was relocated to a fairing at the base of the rudder. This feature was not incorporated on production aircraft.

Su-9 automatic flight control system testbed

Following instructions from GKAT, in 1961 OKB-51 converted a Moscow-built Su-9 (c/n 100000603) into a testbed for the Polyot-1 (Flight-1) automatic navigation/approach/instrument landing system developed for new advanced fighters. The aircraft was equipped with the SAU-1I automatic flight control system (*sistema avtomaticheskovo oopravleniya*); LI and GK NII VVS pilots tested the system in various flight modes, including automatic landing approach and low-level flight, until 1964 (some sources say 1968).

Su-9 target simulation aircraft

During the trials of the Su-15T a single Su-9 was used to determine the efficiency of the new interceptor's Taifoon fire control radar. To this end an angle reflector was installed on the rear fuselage to increase the aircraft's radar cross-section (RCS) for the purpose of emulating a larger aircraft.

T-431 record-breaking aircraft

At an early stage it became apparent that the Su-9's high performance (the estimated service ceiling and maximum speed were 20,000 m/ 65,620 ft and Mach 2.1 respectively) made an attempt on the existing speed and altitude records possible. The Sukhoi OKB decided to have a go at it, and the T43-1 prototype was assigned for the record-setting programme. M. I. Zooyev was appointed engineer in charge of the programme.

On 14th July 1959, after a series of training flights test pilot Vladimir S. Ilyushin established an absolute world altitude record, reaching 28,852 m (94,658 ft) in a zoom climb; this record was later officially recognised by the FAI (*Fédération Aéronautique Internationale*). In the official documents submitted to the FAI the aircraft was referred to as the T-431 (which was very close to the truth), while the powerplant was enigmatically referred to as 'Type 31' – which was also perfectly true, since the AL-7F-1's product code was *izdeliye* 31.

Further record attempts had to be put on hold because of the need to test and refine the TsD-30 radar. The programme resumed in 1962, by which time the T43-1 had undergone major changes. A new and more powerful AL-7F-2 engine was installed, some equipment items (notably in the weapons control system) were removed to save weight, and most of the cooling air scoops on the rear fuselage were likewise deleted to cut drag. On

Above: Su-9 '61 Blue' (c/n 100000610) was converted into the 100L aerodynamics research aircraft for testing possible wing designs for the T-4 missile strike aircraft. This is the first configuration designated 100L-1.

Right: The same aircraft in later configuration as the 100L-2M with mildly ogival wings. The starboard wing is covered in wool tufts for airflow visualisation.

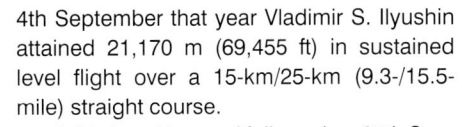

4th September that year Vladimir S. Ilyushin attained 21,170 m (69,455 ft) in sustained level flight over a 15-km/25-km (9.3-/15.5-mile) straight course.

A third world record followed on 25th September 1959 when OKB test pilot A. A. Koznov averaged 2,337 km/h (1,451.5 mph) in level flight over a 500-km (310.5-mile) closed circuit.

T-405 record-breaking aircraft

A further Su-9 – this time a production aircraft (c/n 0405) – was modified in 1960 for the purpose of setting new world records. On 28th May 1960 GK NII VVS test pilot B. M. Andrianov established a world speed record on a 100-km (62-mile) distance, attaining 2,092 km/h (1,299.3 mph). The official documents submitted to the FAI referred to the aircraft by the fictitious designation T-405 derived from the construction number, while the engine was euphemistically called 'Type 13'.

'100L' aerodynamics research aircraft ('100L1' through '100L8')

In 1966-67 a single Moscow-built Su-9 coded '61 Blue' (c/n 100000610) was extensively modified for aerodynamics research as part of the programme to create the T-4 strategic supersonic bomber. Since the latter had the product code *izdeliye* 100, the research aircraft was designated '100L', the letter standing for [*letayuschchaya*] *laboratoriya* – lit. 'flying laboratory'. This Russian term is used indiscriminately for all sorts of testbeds and research or survey aircraft.

The '100L' was intended for exploring the aerodynamics of low aspect ratio wings with a thin, sharply swept leading edge over a wide range of angles of attack (AOAs) and at speeds ranging from Mach 0.3 to Mach 2.0. It was, in effect, a subscale demonstrator helping the engineers to choose the optimum planform for the heavy T-4's wings. The thin wings with a sharp leading edge selected for one of the T-4's project versions were to combine a high lift/drag ratio in supersonic cruise with acceptable field performance.

The conversion involved rewinging a production Su-9 – well, not exactly: the greater part of the wing structure remained unchanged. The leading edge portion of each wing ahead of the main spar was replaced with a new one featuring greater chord and a more pointed profile; as a result, leading-edge sweep was increased from 60° to 65°

Another aspect of the 100L-1.

Above: Su-9 '93' (c/n 1215393) in standard configuration prior to being modified as the L.02-10 control configured vehicle.

Above and below: The L.02-10 as originally modified with two all-movable 'foretails' ahead of the cockpit featuring anti-flutterbooms. The aircraft was never flown in this configuration (with both 'foretails').

The L.02-10 hangared for modifications which involved removal of the upper 'foretail'. The letters on the badge on the nose read KME (*Komsomol'sko-molodyozhnyy ekipazh* – Young Communist League crew).

and wing area increased accordingly. Several modified leading edge sections with different airfoils were fitted consecutively in the course of the tests.

This was not the only change effected on the '100L'. About 1,000 kg (2,200 lb) of ballast were installed in the forward fuselage to shift the CG forward; hence the nose landing gear unit had to be reinforced by fitting a larger nosewheel. The wing tanks were rendered inoperative, reducing the fuel supply by more than 600 litres (132 Imp gal), and converted into test equipment bays; more test equipment was housed in the avionics bays fore and aft of the cockpit. Interestingly, to visualise the airflow over the modified wings the aircraft was equipped with a smoke generator, the smoke exiting via special perforations in the wing leading edge. The modifications increased the aircraft's empty weight by 1,300 kg (2,865 lb) and the maximum take-off weight by 750 kg (1,650 lb) as compared to a stock production Su-9.

The '100L' research aircraft entered flight test in July 1966, the test programme continuing until 1972; LII test pilot E. I. Knyaginichev performed the greater part of the programme. The aircraft was flown in eight different configurations, each one being allocated a sequence number as a suffix to the designation ('100L1' through '100L8'). Three wing versions with a sharp leading edge and one with a rounded leading edge were tested, as were different wing sweep angles and modified tailplanes with a sharp leading edge. Tests revealed that top speed remained unchanged but the landing speed was reduced by a sizeable 40 km/h (nearly 25 mph).

A second Su-9 (c/n 1301) was also converted into a similar research aircraft likewise designated '100L'. The test results gave a wealth of valuable data which allowed the OKB to select the most efficient wing design (a cranked delta) for the T-4 bomber.

L.02-10 CCV

In 1968-75 LII undertook a massive test and development effort with a Su-9 coded '93 Blue' (c/n 1215393) which had been converted to a control configured vehicle (CCV). Canard foreplanes are a familiar thing, but have you ever heard of a canard *foretail*? Well, Su-9 c/n 1215393 featured two all-movable vertical control surfaces of trapezoidal planform installed above and below the fuselage ahead of the cockpit. Actually the upper 'foretail' was found to impair cockpit visibility unacceptably and was removed, all flights being made with only the lower 'foretail' in place.

Originally '93 Blue' was used to investigate the flight dynamics and stability of a directionally unstable aircraft and test an

The L.02-10 as actually flown; the axle of the removed upper 'foretail' protrudes ahead of the cockpit like a sight. Note the forward-looking camera in the fairing at the top of the fin and the non-standard rod aerial for the communications radio aft of the cockpit replacing the usual fin top antenna.

automatic directional stability augmentation system. In 1975-78 the aircraft underwent more modifications performed jointly by LII and OKB-51, receiving a direct side-force control (DSFC) system and the designation L.02-10 (the L stood for [*letayuschchaya*] *lab-oratoriya*). In this guise the aircraft was tested at Zhukovskiy in 1979-84.

U-43 combat trainer prototypes

The aforementioned CofM directive of 16th April 1958 concerning the T-3-51 weapons system also ordered the development of a two-seat combat-capable conversion trainer variant of the T-3. That year, however, OKB-51 was physically unable to commence work on the trainer, being much too busy getting the single-seater up to scratch. Hence on 18th March 1959 the Council of Ministers issued another directive, this time dealing specifically with a trainer version of the T-3 (*sic*). The use of the T-3 designation in this document is rather surprising, since the T-43 interceptor was already undergoing State acceptance trials at the time.

At OKB-51 the aircraft was known as the U-43, implying that it was a trainer derivative of the T-43 (U = *oochebnyy* [*samolyot*] – trainer). To accommodate a second cockpit for the instructor pilot a 600-mm (1 ft 11⅝ in) 'plug' was inserted aft of the cockpit to avoid encroaching on the fuselage fuel tanks; thus the fuel capacity of the single-seat and two-seat versions was identical. The trainee and the instructor pilot sat in tandem under a common canopy featuring individual aft-hinged portions above the two cockpits, with a rigidly installed section in between; the trainee's (forward) canopy section was rather longer than the instructor's.

The U-43 retained the single-seater's avionics suite, including the radar and the Lazoor' GCI data link receiver; radar displays were installed in both cockpits. The ordnance load, however, was reduced to two RS-2-US AAMs, to which the Air Force agreed. In flight the crew members communicated by means of an SPU-2 intercom (*samolyotnoye peregovornoye oostroystvo*). The changes increased empty weight by 630 kg (1,390 lb).

Detail design of the U-43 was completed in the spring of 1960. One of the OKB's development aircraft, a Moscow-built Su-9 known as the T43-14, was selected for conversion into the trainer prototype, the conversion job continuing until late September 1960. The prototype featured slightly reduced fuel tankage (3,430 litres/7554.6 Imp gal) because the wing roots were occupied by test equipment.

On 23rd November, after a series of tests to determine the airframe's resonance frequencies, the U-43 was delivered to Zhukovskiy; K. N. Strekalov was appointed engineer in charge of the flight test programme. Foul weather made the first flight impossible for two months, only taxying trials and high-speed runs being performed in the meantime. On 25th January 1961 the trainer finally took to the air with OKB-51 test pilot Ye. K. Kukushev at the controls. The manufacturer's flight tests proceeded rapidly, poor visibility from the instructor's cockpit being the only major deficiency detected in the course of the five-month programme.

In early May the U-43 was officially submitted for State acceptance trials; however, on 14th July an engine failure occurred when the aircraft was due to be ferried to Vladimirovka AB and the prototype was grounded for an engine change, which delayed the beginning of the trials by another six weeks. It was not until 1st September that GK NII VVS test pilots V. M. Andreyev and Igor' I. Lesnikov started flying the *sparka*. (This slang appellation for a trainer is derived from *sparennoye opravleniye* – dual controls – and translates loosely as 'Two Sticks'.) B. M. Korshunov was appointed engineer in charge of the U-43's trials, with Ye. N. Boobnov as his assistant.

The trials programme was completed on 23rd December, involving a total of 83 flights. With two K-5M (RS-2-US) AAMs the trainer's acceleration parameters and service ceiling were almost identical to those of the single-seat Su-9 armed with four missiles. At 12,800 m (42,000 ft) the trainer attained its V_{NE} of 2,230 km/h (1,385 mph) or Mach 2.1 in full afterburner. At an all-up weight of 9,700 kg (21,380 lb) with missiles and drop tanks the service ceiling in level flight was 17,500 m (57,415 ft); range with full external stores at 11,000 m (36,090 ft) was 1,370 km (850 miles). The aircraft possessed all-weather, day/night capability, making it possible to perform conversion training and train pilots in intercepting targets flying at speeds of 800-1,600 km/h (495-990 mph) and altitudes up to 20,000 m (65,620 ft); the interception range determined by available fuel was 105-130 km (65-80 miles).

The State acceptance trials protocol said that the two-seater met all of the customer's requirements, except for the inadequate

Top and above: The U-43 (or possibly U43-1) – the prototype of the Su-9U trainer. These views illustrate the trainer's longer forward fuselage and wheelbase. Note the vertical canopy frame member on the trainee's canopy section, a characteristic feature of the prototype, and the wider windscreen.

visibility from the rear cockpit. The State commission recommended the U-43 for production and service, providing this and a few other shortcomings were eliminated.

Hence the prototype was flown back to Moscow for modification work which continued until February 1962. The headrest of the trainee's ejection seat was cropped at the sides and the glazing area of the forward canopy section was increased for better visibility; in the rear cockpit, a section of the instrument panel was moved to the right and the glass pane separating the two cockpits was widened. On 23rd March Ye. K. Kukushev reflew the upgraded prototype; a brief checkout test programme at GK NII VVS followed on 11th-26th April and the military were satisfied with the field of view from the rear cockpit.

(Note: Some sources claim that two trainer prototypes were built and that the changes described above were incorporated on the second prototype (U43-2), whereas the first aircraft (U43-1) remained unmodified. As originally flown with the old canopy design the U-43 had star insignia, whereas a photo of a prototype with a revised canopy shows no insignia whatever; there is no logical reason why the insignia should have been removed in the process of conversion.)

Su-9U combat trainer (*izdeliye* 11)

Back in 1961, having received a set of manufacturing documents from the Sukhoi OKB, MMZ No.30 at Moscow-Khodynka built an initial-production batch of U-43 trainers. The aircraft received the service designation Su-9U and the factory code *izdeliye* 11 (since it followed the single-seater version which, as the reader remembers, was *izdeliye* 10). Only 50 examples had been completed when production ended in 1962. Unlike the single-seater, the Su-9U featured an AP-28Zh-1 autopilot, a D-3K-110 yaw damper and red cockpit lighting from the outset. Production aircraft had longer range than the prototype thanks to the wing tanks (on the U-43 they were eliminated to provide room for test equipment).

L-43 ejection seat testbed (*izdeliye* 94)

Upon completion of the State acceptance trials the U-43 prototype (according to some sources, the first prototype) was converted into an ejection seat testbed for verifying new ejection seats developed by the Sukhoi OKB. The modified aircraft received the designation L-43 or *izdeliye* 94. The tests confirmed that the latest seats in the KS series permitted safe ejection at speeds up to Mach 1.8 and altitudes of 150-15,000 m (490-49,210 ft).

According to some sources, in 1962 the first two production Su-9Us (c/ns 1110000101 and 111000102?) were converted for testing new ejection seats and high-altitude pilot gear, the modified rear cockpits permitting installation of seats developed by various manufacturers. One aircraft was delivered to LII, operating from Zhukovskiy, while the other belonged to GK NII VVS and was home-based at Vladimirovka AB. On the other hand, and rather confusingly, there is evidence that in 1967-75 LII operated 'Su-9U c/n 1018' (ie, c/n 112001018?) converted into an ejection seat testbed which was used for State acceptance trials of the Sukhoi KS-1, Mikoyan KM-1 and Yakovlev KYa-1 seats, as well as the Czech VS-1BRI zero-zero ejection seat developed for the Aero L-39 Albatros advanced trainer (VS = *vystrelovací sedacka* – ejection seat) and the later VS-2 model. The exact number and identities of the Su-9U ejection seat testbeds is unknown (different sources give conflicting data), but one aircraft was coded '10 Blue'.

Top: Some sources suggest this aircraft totally devoid of markings is the second prototype Su-9U (U43-2).
Above: Probably the same aircraft after the insignia had been applied. Note the revised canopy adopted for the production version.

The changes made to the aircraft were as follows. The pressurised cockpit was divided into two separate pressurised compartments, making sure that the forward cockpit remained pressurised and the pilot thus remained able to fly the aircraft after the rear seat had been fired. The flight controls, engine controls and instruments were deleted from the rear cockpit; the normal canopy could be fitted for ferry flights but was removed for test flights and a special open-top fairing was fitted instead. The testbed was equipped with three high-speed cine cameras to capture the ejection sequence – two in fairings on the wings and one in the cockpit. Performance and handling were almost identical to those of the standard trainer.

Ejections were performed during the take-off run, at high altitude and at supersonic speeds; initially test dummies were used, of

One of the L-43 ejection seat testbeds featuring a modified rear cockpit and photo calibration markings on the fuselage and tail. Note that the canopy of the forward cockpit is identical to that of the prototype as originally flown.

Above: An ejection seat is fired from the L-43 testbed, now coded '10 Blue'.

A Zvezda K-36 ejection seat is fired from the same aircraft flying at high altitude.

T-3 series aircraft production at plant No.153

Manufacturer's designation	Product code	Construction number	Notes
PT-7	Izdeliye 27	0115301 through 0115307 (T1-01 through T1-07)	Wing dogtooth
PT-7	Izdeliye 27	0115308 through 0215309 (T1-08 through T2-09)	Wing dogtooth; some aircraft between c/ns T2-02 and T2-09 converted to PT-8s
PT-8	Izdeliye 27	0215302 through 0215310 (T2-02 through T2-10)	Wing dogtooth
T-43 †	Izdeliye 27	0215310 through 0315310 (T2-10 through T3-10)	Wing dogtooth
T-43 †	Izdeliye 27	0315311 through 06153… (T2-10 to end of Batch T6)	No wing dogtooth
T-43 †	Izdeliye 34	0715301 through 1515350 * (T7-01 through T15-50)	No wing dogtooth
T-47 ‡	Izdeliye 36	1015351 through 1515350? (T10-51 through T15-50?)	No wing dogtooth

* Aircraft with c/ns 0715301 through 1015351 (T7-01 through T10-51) had sloping fuselage frames Nos 17-19 and provisions for installing NR-30 cannons in the wing roots.
† The Su-9 (T-43/*izdeliye* 27) had flexible fuel cells, while the Su-9 (T-43/*izdeliye* 34) had integral tanks.
‡ The T-47 is described in Chapter 2.

course, but these were later joined by parachutists made of flesh and blood. Later the Su-9U ejection seat testbeds were used for testing the Zvezda K-36 ejection seat which has become a standard fit on all Soviet/Russian combat aircraft developed from the late 1970s onwards.

Su-9U control system testbed

In 1962-64 LII used a modified Su-9U (identity unknown) to test an automatic stability augmentation system.

L.07-10 aerodynamics research aircraft

In 1975-76 the Sukhoi OKB extensively modified a Su-9U (c/n 112001301) as part of the effort to develop a fourth-generation fighter known in-house as the T-10 (NATO codename *Flanker-A*) – ie, the precursor of the famous Su-27. Designated L.07-10 (ie, 'flying laboratory' No.7 under the T-10 programme?), this aircraft was intended for exploring the T-10's wing aerodynamics; it featured ogival wings wings with a sharp leading edge having compound curvature. The L.07-10 entered flight tests at LII in 1977; tragically, it crashed after suffering a birdstrike on take-off in 1982, killing the pilot.

To conclude this chapter the author would like to give some data on the production of T-3 series aircraft at plant No.153 in Novosibirsk. (Interestingly, the T-3 designation persisted for a long time, appearing in the manufacturing documents long after the pure T-3 had vanished.) It should be noted that in the case of this aircraft family the phrase 'production aircraft' need not always be taken literally because, even though some machines were built by the series production factory, to all intents and purposes they were prototypes. (Indeed, at the initial stage of production almost all 'production' aircraft were in fact prototypes of some sort!) Also, remember that the OKB and the factory sometimes used different in-house codes for the same aircraft.

As already mentioned, the full construction numbers included the factory number placed between the batch number and the number of the aircraft within the batch (for instance, 0315304); however, on the actual aircraft the c/n was stencilled on the main landing gear doors in an abbreviated form and with a dash (03-04). Furthermore, the factory had a habit of putting the letter T (for *treugol'noye krylo* – delta wings) before the batch number in internal documents and omitting the zero in batches 01 through 09. Thus, 'T1-07' does not mean the seventh prototype T-1 (this aircraft, as we remember, was never built) but denotes T-3 c/n 0115307. Similarly, 'T10-51' is Su-11 c/n 1015351, not a Su-27 (T-10S) development aircraft.

Chapter 2
The Flying Pipes

T-47 experimental interceptor (T47-1 prototype)

When the Sukhoi OKB started work on equipping the T-3 interceptor with a powerful Almaz fire control radar, the resulting aircraft received the manufacturer's designation T-47. Again, the radar featured separate search and tracking antennas. On the T-47, however, the airframers managed to reach a compromise with the radar makers and reconcile good aerodynamics with acceptable radar performance by placing both antennas inside a large conical centrebody within an axisymmetrical air intake. To accomplish this the forward fuselage diameter was significantly increased and two large dielectric panels were incorporated on the sides of the nose to ensure an acceptable directional pattern for the radar's search antenna.

Of course, the intake aerodynamics and cockpit visibility deteriorated somewhat as a result of this redesign. However, this was considered an acceptable trade-off because the nation urgently needed a high-altitude interceptor to neutralise the threat posed by high-flying American spyplanes. The situation called for an immediate decision on what armament the future interceptor was to carry. In addition to air-to-air missiles, the installation of Nudelman/Rikhter NR-30 cannons and unguided rocket pods was still considered at this stage.

On 18th December 1957 GKAT and the Soviet Air Force issued a joint ruling requiring the Novosibirsk aircraft factory No.153 to manufacture an initial batch of ten T-47 interceptors in 1958. The aircraft were to be equipped with the Almaz radar featuring co-located search and tracking antennas in a movable intake centrebody (shock cone) and be armed with NR-30 cannons; no missiles were envisaged in the first batch.

Construction of the first prototype T-47 (known as the T47-1) proceeded in accordance with an as-yet-MAP order issued on 6th August 1957. To this end the second production PT-8 (c/n 0115302) was delivered to OKB-51's prototype construction shop in Moscow in the form of separate subassemblies. After appropriate modifications to the fuselage nose the aircraft was fitted with the Almaz radar, two cannons in the wing roots and two ORO-57 rocket pods for 57-mm (2.24-in) ARS-57 folding-fin aircraft rockets. On 10th December 1957 the aircraft was delivered to the OKB's flight test facility in Zhukovskiy, V. I. Mosolov being assigned as engineer in charge of the test programme.

On 6th January 1958 the T47-1 performed its maiden flight. Soon, however, the manufacturer's flight tests had to be suspended because another development aircraft, the PT8-4 (see below) was due to enter flight test imminently; AL-7F-1 engines were still in short supply then, so the engine was simply removed from the T47-1 and installed in the PT8-4. Hence by early June 1958 the T47-1 had made a mere 15 flights – and that was it. On 4th June 1958 the Council of Ministers issued directive No.608-293 ordering all work on the K-7 weapons system and the cannon-armed T-47 to be stopped and all efforts to be directed towards developing more advanced weapons systems.

Actually this was not yet the end of the T47-1's flying career. In September 1958 the aircraft was transferred from the Sukhoi OKB to the Flight Research Institute; this establishment used it until 1963 as a propulsion testbed for refining the AL-7F-1 engine.

PT8-4 experimental interceptor

The work on arming the prospective high-altitude interceptor featuring the Almaz radar with air-to-air missiles enjoyed priority over the cannon/rocket-armed T-47 which had entered low-rate initial production (LRIP) by then. Hence in keeping with MAP order No.718 of 19th November 1957 in the autumn of 1957 OKB-51 commenced conversion of the first production PT-8 (c/n 0115301) delivered from Novosibirsk. Since this was the fourth example built there, being preceded by the three pre-production PT-8s in batch 00, the aircraft picked for conversion into the missile-armed variant was designated PT8-4.

Like the T-47 development aircraft described above, the PT8-4 was modified by the OKB's experimental shop to feature a dogtooth wing leading edge and reduced-area ailerons; the fuselage remained unaltered for the time being. The aircraft was prepared for initial flight testing in production-standard PT-8 configuration with separate radomes for the search and tracking antennas, a narrow aft fuselage and an AL-7F

Top and above: The PT8-4 prototype, showing the dielectric inset panels on the sides of the nose, the increased-diameter rear fuselage housing an AL-7F-1 engine and the wing leading edge dogtooth.

27

This view of the PT8-4 illustrates to advantage the new constant-diameter nose with the much-enlarged shock cone and the cannons in the wing roots.

engine. K. K. Solov'yov was appointed engineer in charge of the test programme.

However, drawing on the first results of the T43-1 prototype's tests which had just begun, in late October 1957 it was decided to modify the PT8-4 to T-47 standard as regards fuselage design. The second conversion involved replacing the fuselage nose ahead of the cockpit with a new structure featuring a larger-diameter axisymmetrical air intake and shock cone to take the Almaz-3 radar and manufacturing a new, wider detachable rear fuselage section designed to accommodate the AL-7F-1 (*izdeliye* 45-1) engine. Wing pylons and the appropriate controls were installed for carrying and launching K-7L or K-6V AAMs, and a pair of NR-30 cannons was also installed in the wing roots. Thus the T-47 and the PT8-4 were the only aircraft of the T-3 family to feature cannon armament.

Initially the PT8-4 was fitted with an AL-7F engine because the intended AL-7F-1 was still unavailable. Following completion in January 1958 the aircraft was trucked to Zhukovskiy in the closing days of the month to begin ground checks. On 21st February it became airborne for the first time with Sukhoi OKB test pilot Vladimir S. Ilyushin at the controls.

Initiated by GKAT order No.49 of 15th February 1958, the manufacturer's flight tests of the PT8-4 proceeded in accordance with the so-called Programme 1, which involved checking the aerodynamics and performance/handling characteristics of the aircraft with the redesigned forward fuselage and wing dogtooth. The first five flights under Programme 1 showed that a special test programme was necessary to explore certain peculiarities of the aircraft. Such a programme (the so-called Programme 2) was duly drawn up and completed, whereupon Programme 1 resumed.

By the end of June 1958 the PT8-4 had made a total of 33 flights, including eight under Programme 2; four versions of the air intake centrebody were tried out and the AL-7F-1 engine was finally installed. Thus the T-47 configuration of the T-3 interceptor had, in effect, been put through its paces on the PT8-4.

When giving their recommendations concerning wing design, the specialists at TsAGI were effectively staging a large-scale flight experiment to see which way of preventing tip stall (boundary layer fences, as used on the Mikoyan/Gurevich Ye-5 and Ye-6, or a leading-edge dogtooth as used on the Sukhoi T-3) was more effective. The dogtooth showed disappointing performance during the tests of the T47-1 and the PT8-4, and all subsequent aircraft of the T-3 family reverted to an unbroken wing leading edge.

During manufacturer's flight tests the T8-4 displayed the following performance. The take-off weight was 8,960 kg (19,750 lb) without missiles and 9,330 kg (20,570 lb) with two K-7 missiles. Top speed at 13,000 m (42,650 ft) in full afterburner was 2,260 km/h (1,403 mph) in 'clean' configuration (without missiles), decreasing to 2,180 km/h (1,354 mph) with two AAMs. At altitudes up to 8,000 m (26,250 ft) the top speed was restricted to 1,250 km/h (776 mph) by a dynamic pressure limit of 7,500 kg/m² (1,537 lb/sq.ft); between 8,000 m and 10,000 m (32,810 ft) it was 1,200 km/h (745 mph) restricted by a dynamic pressure limit of 7,000 kg/m² (2,135 lb/sq.ft).

The maximum permissible Mach number at 11,000 m (36,090 ft) and higher was 2.1. In 'clean' configuration the aircraft could reach 1,250 km/h indicated airspeed but was not to exceed Mach 2.25; with two missiles the limit was still 1,250 km/h IAS but the maximum Mach number was 1.19.

The service ceiling at full military power was 15,000 m (49,210 ft). The service ceiling in full afterburner was never established due to the premature termination of the tests. The aircraft did, however, reach a maximum altitude of 19,000 m (62,335 ft) with the ultimate (No.4) air intake centrebody in the fully aft position, climbing at Mach 1.6. The PT8-4's maximum landing weight was 8,305 kg (18,310 lb) and the G limit was set at +7 Gs.

Apart from V. S. Ilyushin, the aircraft was flown by OKB test pilots A. A. Koznov and V. N. Il'yin, as well as by the Novosibirsk aircraft factory's checkout pilots. The PT8-4's flying career was not altogether accident-free. On 5th April 1958, when the aircraft was making its second flight with the AL-7F-1, the engine surged and quit. After several fruitless attempts to relight the engine Ilyushin, in a remarkable display of airmanship, glided the aircraft back to base and performed a perfect dead-stick landing – the first successful dead-stick landing in an aircraft of the T-3 family.

As already mentioned, the CofM directive of 4th June 1958 which killed the cannon-armed T-47 also terminated further development of the K-7 missile in favour of new weapons systems. Shortly afterwards, in early August 1958, the PT8-4's career came to an abrupt end. An electrics failure in one of the test flights put the artificial feel mechanism in the tailplane control circuit out of action, forcing test pilot Eduard V. Yelian to make a landing at an excessively high approach speed. To top it all, the brake parachute failed to deploy and the aircraft overran, suffering damage so grave that it was declared a write-off – for the time being, as it turned out. Later the PT8-4 was extensively rebuilt, emerging as the T47-3 interceptor prototype – but that's another story (see below).

T-3-8M aerial intercept weapons system

T47-2, T47-3, T47-4, T47-5, T47-7 and T47-8 interceptor fighter prototypes

Several other interceptors bearing the T-47 designation but equipped with different radars were evolved from the T-3 at OKB-51. The Oryol (Eagle) radar developed by OKB-339 under G. M. Koonyavskiy as a refined derivative of the production RP-6 *Sokol* (Falcon) fitted to the Yakovlev Yak-25 and Yak-27 interceptors had greater appeal than the other contenders – first and foremost because it featured a single antenna. On the Yakovlev types the Sokol radar worked with the K-8 AAM developed by OKB-4 under Matus Ruvimovich Bisnovat; this advanced missile had much greater range than the K-5M and, importantly, came in both semi-active radar homing (SARH) and infra-red homing versions.

When Aleksandr S. Yakovlev's OKB-115 ran into development problems with the Yak-27K interceptor, the Ministry of Aircraft Industry proposed adapting the Yakovlev twinjet's armament and fire control radar to the Sukhoi T-3; as already mentioned in the previous chapter, a Council of Ministers directive to this effect appeared on 16th April 1958. OKB-4 was tasked with developing a 'Sukhoi version' of the missile designated K-8-2 (or K-8M), while OKB-339 was instructed to reduce the RP-6 Sokol radar's huge antenna dish to a size small enough to fit inside the air intake shock cone used by the Almaz radar. The K-8M missile was heavier than the K-5MS, having a launch weight of 225 kg (496 lb) and a 35-kg (77-lb) warhead. Unlike the initial K-8 developed for the Yak-27K, which featured interchangeable IR seeker heads with different sensitivity for day and night use, the 'Sukhoi version' was developed in one 'round-the-clock' version. Thus, even though target engagement was still possible only in pursuit mode, the new radar and armament gave the aircraft a considerably higher potential as compared to the T-43 prototype then undergoing trials.

Detail design of the interceptor featuring the Oryol radar and K-8M missiles began in the second half of 1958. All the aircraft made available by the termination of the K-7 weapons system – the PT-7 prototype, the wrecked PT8-4 and several LRIP examples of the T-47 – were earmarked for conversion to this configuration; a total of six aircraft was to be tested. The first of these to enter flight test was the T47-2; it was virtually unmodified and no radar was fitted, since the aircraft was purely an aerodynamics test vehicle. Soon afterwards, however, the T47-2 crashed and was destroyed; the pilot survived.

The first example to be converted under the T-3-8M programme (ie, T-3 armed with K-8M missiles) was the PT-7. In the summer of 1958 the aircraft was flown to Moscow and underwent extensive modifications at the Sukhoi OKB's prototype construction shop, receiving a T-47 style nose and a new designation, T47-3. The air intake shock cone was of all-metal construction, not dielectric, since the intended radar was still unavailable; nevertheless, during the initial stage of the trials the OKB intended to use the T47-3 not only for aerodynamic testing but for verifying the missile launch system as well (apparently the IR-homing version of the missile was used).

On the night of 26th November 1958 the T47-3 was trucked to the Sukhoi OKB's flight test facility in Zhukovskiy. Manufacturer's flight tests began on 25th December, with R. G. Yarmarkov as engineer in charge. Initially the aircraft was used to verify the intake design and evolve the optimum intake adjustment algorithm; the shock cone and auxiliary blow-in doors were adjusted manually for the time being, as automatic controls had yet to be developed. The data obtained at this stage allowed the parameters of the new ESUV-2 intake control system developed for the T-3 with the new air intake to be determined. The early test results obtained with the T47-3 confirmed that the redesigned forward fuselage adversely affected the flight performance, causing a deterioration of the acceleration parameters, a reduction of the service ceiling, top speed and range.

On 20th October 1958 the first fatal crash of a T-3 series aircraft occurred at Novosibirsk-Yel'tsovka, the factory airfield. When factory test pilot V. V. Proveshchayev was making a checkout flight in a production fighter completed to T-47 standard the radome suddenly disintegrated, damaging the engine, which quit. After several unsuccessful attempts to restart the engine Proveshchayev attempted an off-field forced landing but was killed in so doing.

Two more aircraft converted from T-3s – or, to be precise, LRIP T-47s – at the Novosibirsk aircraft factory joined the test programme in the spring of 1959. The T47-4 converted in Moscow from T-47 c/n 0115309 arrived at the LII airfield in Zhukovskiy in April and V. Vasil'yev was assigned to this aircraft as engineer in charge of the tests; in May it was followed by the T47-5 (formerly T-47 c/n 0215302), with V. Balooyev as engineer in charge. These two aircraft were earmarked for armament/weapons control system trials – specifically, for testing the IR-homing version of the K-8M. The T47-5 featured the Oryol radar from the start, while the other aircraft initially had none, being retrofitted in the course of modifications in January 1960. In August 1959, having passed pre-delivery checkout tests, both fighters were flown to the GK NII VVS facility at Vladimirovka AB to commence manufacturer's tests of the radar and the K-8M missile. These were performed by OKB-51 test pilots Vladimir S. Ilyushin, Yevgeniy S. Solov'yov, Ye. K. Kukushev and A. A. Koznov who made ten missile launches in radar-guided mode and about 40 flights to check the radar's operation during intercepts of real targets. By mid-September the greater part of the programme had been completed, and on 17th September 1959 the T-3-8M aerial intercept weapons system was officially submitted for State acceptance trials.

Stage A of the trials lasted from November 1959 to April 1960, involving trials of the aircraft with the IR-homing version of the missile according to an abbreviated programme agreed upon with the Air Force. Three of the prototypes (the T47-3, T47-4 and T47-5) participated in this stage, flown by GK NII VVS pilots N. P. Zakharov, P. F. Kabrelev, Boris M. Andrianov, E. I. Knyazev and V. M. Andreyev, as well as virtually all Sukhoi OKB test pilots. By the time this stage ended two more prototypes (the T47-7 and T47-8) were ready to join the action. These two aircraft were earmarked for testing the SARH version of the K-8M, and the latter example incorporated major changes made at the request of the military. Specifically, the T47-8 was powered by an uprated AL-7F-2 delivering 10,100 kgp (22,270 lbst) in full afterburner. An extra fuel tank was provided in the forward fuselage and the existing bag-type rear fuselage fuel tank

The T47-4 prototype carrying two K-8M missiles. The darker shade of the metal skin ahead of the cockpit windscreen shows clearly that the nose is a new structure grafted onto an airframe built with a different intake design. Note the 'towel rail' aerial aft of the nose gear unit.

gave place to an integral tank; also, the dry bays between the wings' main spar and No.1 auxiliary spar were transformed into integral tanks. The combined effect of these changes was to increase the internal fuel capacity to 4,195 litres (922.9 Imp gal) and the fuel load to 3,440 kg (7,580 lb). Finally, two detachable wiring conduits were installed on the upper centre fuselage sides to facilitate maintenance access.

On 26th April 1960 all five aircraft commenced Stage B of the State acceptance trials which was mostly concerned with the verification and debugging of the Oryol radar and development of operational tactics (the latter involved live missile launches). Additionally, the shape of the radome/shock cone was optimised and the new ESUV-2 automatic air intake control system was put through its paces. By the middle of the year the T47-3, T47-4, T47-5 and T47-7 had been upgraded to match the standard of the T47-8. Trials showed that the extra fuel increased the range to an acceptable figure.

The trials were completed in late May 1961; on 8th June the State commission signed the final protocol. In the course of the State acceptance trials the five aircraft involved made a total of 475 flights; together with the manufacturer's flight tests the number of flights exceeded 700.

On 9th July 1961 the new interceptor had its public debut when the T47-8 flown by OKB test pilot Yevgeniy S. Solov'yov participated in the traditional Aviation Day flypast at Moscow-Tushino.

T47-4 development aircraft
Later, when the State acceptance trials of the T-3-8M aerial intercept weapons system had been completed, the T47-4 was fitted with an AP-28E-1 autopilot and an ESUV-1 electrohydraulic air intake control system, serving as a testbed for these systems.

Still later, the T47-4 served as a testbed for some components of the SAU-58 automatic flight control system created by the Sukhoi OKB for the T-58 interceptor (the future Su-15, see Chapter 5).

T47-5 development aircraft
During the same period (after the State acceptance trials of the T-3-8M weapons system) the T47-5 was used as a weapons testbed for the IR-homing version of the K-8M air-to-air missile.

T47-6 experimental interceptor
The reader has probably noticed that one of the aircraft in the above sequence is 'missing'. This aircraft, the T47-6, was set aside for conversion under a different programme. Again, the aircraft was converted from a standard LRIP T-47.

Initially the T47-6 acted as an aerodynamics research aircraft, being fitted with canard foreplanes and a modified air intake featuring a fixed shock cone/radome. In this guise the aircraft flew both with an AL-7F engine and with an AL-7F-1.

Later the T47-6 was used in the development of the T-3A-9 aerial intercept weapons system and the ill-starred Sukhoi T-37 heavy interceptor described separately in this book. The T47-6 acted as a testbed for the TsP-1 fire control radar and the launch control system for the K-9 (R-38) SARH air-to-air missiles, both of which had been created for the T-37. Again, the aircraft was converted from a standard LRIP T-47, featuring the T-37's fixed multi-shock intake centrebody/radome and PR-38 missile launch rails under the wings.

The aircraft entered flight test in January 1960. Several versions of the intake were tried, the final one yielding unique results with absolutely stall-free operation. By then, however, the T-47 interceptor equipped with the Oryol radar and K-8M missiles had entered full-scale production and it was considered inexpedient to make major design changes.

Later, when the T-37 programme had been cancelled, the T47-6 was used for testing the T-3-8M aerial intercept weapons system alongside the other prototypes in the series.

Su-11-8M aerial intercept weapons system

Su-11 interceptor (*izdeliye* 36)
On 5th February 1962 the T-3-8M weapons system was officially included into the Soviet Air Force inventory. In so doing the T-47 interceptor received the service designation Su-11; its Oryol radar was officially designated RP-11 to match the aircraft type, the

Left: Close-up of the forward fuselage of the T47-4, showing the absence of the inset dielectric panels on the nose and the gun blast plates.

Below: The T47-4 wearing the appropriate inscription on the tail and the tactical code '35 Blue' in its days as an instructional airframe at the PVO technical school in Solntsevo just outside Moscow. Note the wiring conduits on the centre fuselage sides.

Above: The T47-5 carrying orange-painted inert K-8M missiles and wearing four 'kill' stars to mark successful missile launches. Note the all-metal intake centrebody.

K-8M AAM was redesignated R-8M (R for *raket*a – in this case, missile) and the system as a whole was redesignated Su-11-8M.

Production was organised at plant No.153 named after V. P. Chkalov in Novosibirsk, the aircraft receiving the in-house product code *izdeliye* 36; a Council of Ministers directive to this effect appeared on 27th November 1961. According to this document the new aircraft was to completely supersede the Su-9 on the Novosibirsk production line by mid-1962 and the plant was required to manufacture 40 aircraft by the end of that year.

The factory started gearing up for production in 1961; no great difficulties were encountered, since the two aircraft had considerable structural and systems commonality. This 'nose job' changed the aircraft's appearance considerably, resulting in a fuselage of almost constant diameter; this gave rise to the nickname *letayushchaya trooba* (flying tube).

Production Su-11s featured an upgraded avionics and equipment suite; this included an RSIU-5V two-way VHF communications radio, a new ARK-10 automatic direction finder, an ARL-S Lazoor' data link receiver working with the Vozdookh-1 GCI system, an MRP-56P marker beacon receiver, SOD-57M distance measuring equipment, an SRZO-2M Khrom-Nikel' (Chromium-Nickel) IFF interrogator/transponder, a Sirena-2 radar warning receiver (RWR), an AGD-1 artificial horizon and a KSI compass system for fighters (*koorsovaya sistema istrebitel'naya*).

The aircraft also featured many systems components not found on the Su-9 (albeit some of them were later retrofitted to the latter type), namely an RV-UM low-range radio altimeter, a D-3K-110 three-channel yaw/pitch/roll damper, an AP-28Zh-1B autopilot, red cockpit lighting making the aircraft less observable to the enemy at night and a new KS-3 ejection seat permitting safe ejection over a much wider range of speeds and altitudes. Early Su-11s featured a 570 x 140 mm KT-100 nosewheel but this was soon replaced by a 600 x 155 mm (23.6 x 6.1 in) KT-104 wheel.

The first production aircraft (c/n 0115301) took to the air in July 1962. Production was gaining momentum and everything seemed to be going nicely when suddenly disaster struck. On 31st October 1962 the engine of Su-11 c/n 0115301 quit during one of the test flights. The aircraft was passing over Novosibirsk at the time and GK NII VVS test pilot V. M. Andre-yev chose not to eject, fearing that the uncontrollable aircraft would drop in a residential area, causing massive destruction and a large loss of life. He attempted a dead-stick landing on the outskirts of the city's old disused airfield and did reach it but the aircraft touched down so hard that the pilot was killed instantly.

This tragic crash dealt a crippling blow to the Su-11. Air Marshal Yevgeniy Yakovlevich Savitskiy, who commanded the IA PVO, was extremely unhappy about the Su-9's high accident rate as it was, and the accident played into the hands of the Su-11's opponents. At the time Aleksandr S. Yakovlev, a man who still had considerable influence in the aircraft industry, was actively promoting his new Yak-28P twinjet supersonic interceptor, stressing its alleged advantages over the single-engined Sukhoi deltas. He soon succeeded in convincing the Powers That Be that future interceptors should have two engines powering separate electric and hydraulic

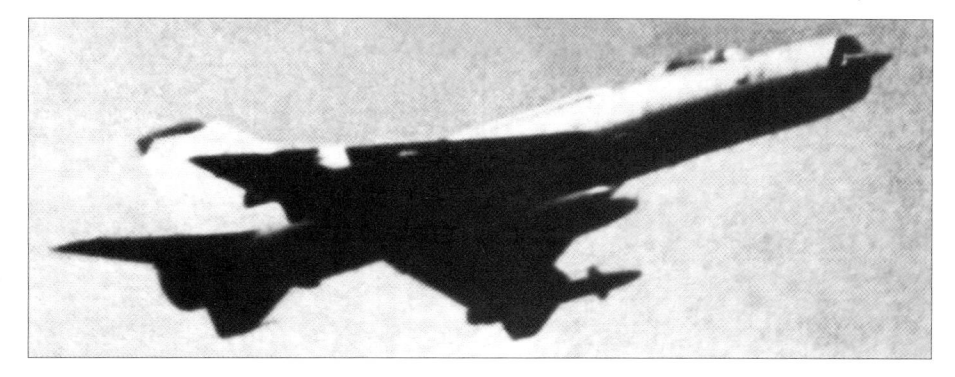

Above and below: The T47-8 pictured during its Tushino appearance on 9th July 1961. The aircraft is carrying two missiles (port, radar-homing, starboard, IR-homing) and two drop tanks.

Above: This Su-11 coded '32 Blue' was a testbed of some sort, as indicated by the absence of the weapons pylons.

systems to ensure greater reliability and operational safety.

As a result, the Novosibirsk aircraft factory's production plan for 1962 came to include 40 Su-11s and 15 Yak-28P two-seat interceptors in addition to the final 120 Su-9s. All subsequent wrangling with the customer which demanded that the Su-11's service reliability be improved by the provision of automated systems monitoring equipment (which, at first glance, was perfectly reasonable) was in reality but a pretext to cut production plans for the Su-11 which had fallen into disfavour at the top level. In 1963 the plant was reoriented towards Yak-28P production (what irony!). By then, however, a considerable stockpile of Su-11 components had been built up; on consideration it was decided not to throw them away but to build fighters as long as the stock lasted, delivering them to the IA PVO after the required modifications had been made. Yet the upgrading process turned out to be a protracted affair, the customer being extremely demanding (or rather biased) after the abovementioned accident, and the completed Su-11s sat on the pre-delivery line at Novosibirsk-Yel'tsovka for weeks. Even when the plant was cranking out and delivering Yak-28Ps (known in-house as *izdeliye* 40) at a steady rate, the factory apron was still crammed with row upon row of Su-11s wrapped in tarpaulins – much to the displeasure of Aleksandr S. Yakovlev – because the PVO top brass could not find the nerve to order them delivered to operational units. At length Yakovlev gave vent to his displeasure during one of his visits to Novosibirsk; pointing to the undelivered Su-11s, he loudly and sourly enquired from the accompanying factory representatives: 'How much longer is this junk going to stay here?'.

In early 1963 the VVS transferred several production Su-11s to OKB-51 in order to have the reliability improvement measures incorporated. Throughout 1963 and 1964 these aircraft underwent extensive testing and modifications. The effort paid off – in mid-1964 the Su-11 finally achieved IOC with the PVO's 393rd GvIAP based near Astrakhan' (*gvardeyskiy istrebitel'nyy aviapolk* – Guards fighter regiment). (The Guards units were the elite of the Soviet armed forces, as this title was accorded to a unit for special gallantry in combat.) By then the Sukhoi OKB was commencing the trials of the new T-58D twinjet interceptor (the future Su-15); this aircraft fitted the new concept ideally, and the OKB placed its bets on the T-58D.

Su-11 production continued until early 1965 but only 108 examples were completed. The remainder of the production run was enough to equip two more PVO regiments of the Moscow PVO District – the 790th IAP at Khotilovo AB and the 191st IAP in Yefremov – in the first six months of 1965.

Despite being built in such small numbers, the Su-11 soldiered on alongside the Su-9 until the early 1980s when the last examples were retired as time-expired. In the late 1960s the Su-11 fleet was upgraded by the addition of SARPP-12 flight data recorders, just as was the case with the Su-9.

Left and above: The tenth production Su-11 ('10 Blue', c/n 0115310) with two heat-seeking R-8M missiles (the starboard one is c/n 2019). The aircraft's c/n is stencilled on the tail as 01-10.

Chapter 3
Defending the Homeland

The Su-9/Su-11 in Service

As early as 1959 the Su-9 achieved initial operational capability with the fighter arm of the Air Defence Force (IA PVO), gradually replacing the subsonic cannon-armed MiG-17PF and the missile-toting supersonic MiG-19PM. Mastering the new interceptor proved to be quite a challenge for flight and ground crews alike, owing to the extensive new technology which had gone into the design of the Sukhoi delta-winged jet and the complexity of its equipment. On the one hand, the Su-9 had a much wider speed and altitude envelope than the predecessors; on the other hand, the aircraft and its systems still suffered from numerous teething troubles. As a result, accident rates skyrocketed during the Su-9's service introduction period, which had a long-term negative effect on the type's service career.

The first production Su-9s were delivered to a PVO fighter regiment stationed at Novosibirsk-Tolmachovo airport, a stone's throw away from aircraft factory No.153 at Novosibirsk-Yel'tsovka, in June 1959. (In Soviet times quite a few airfields served a dual purpose, being both civil airports and Soviet Air Force bases; Novosibirsk-Tolmachovo was one of them.) This was common practice during the service introduction of a new aircraft type; the close proximity of the manufacturer allowed any problems to be speedily solved. Concurrently the PVO's 148th TsBP i PLS located at Savostleyka airbase near Gor'kiy (sometimes rendered as Savasleyka) dispatched a group of highly skilled pilots to the regiment to take conversion training in situ. (Normally the 148th TsBP i PLS would be the first to receive the new hardware, but here the procedure was reversed because of the need to stay close to the factory.) IA PVO Commander Air Marshal Yevgeniy Ya. Savitskiy tasked the unit's personnel with mastering the type as quickly as possible and then participating in the deliveries of brand-new Su-9s to operational units. By early July the PVO pilots had made 72 flights in the first eight production fighters; six service pilots had received their type ratings, with another three in the training pipeline.

When Su-9 production was completed in late 1959 a total of 163 aircraft had been built; more than 150 of these were delivered to first-line units. The abovementioned PVO regiment at Novosibirsk-Tolmachovo was quickly followed by others stationed in Krasnovodsk (Turkmenia, Turkestan Defence District; the city is now renamed Turkmenbashi), Baranovichi (Belorussian DD), Karshi (Uzbekistan, Central Asian DD) and at Kilp-Yavr AB near Murmansk (Leningrad DD).

All aircraft were ferried to the units from Novosibirsk, not delivered in crates by rail, as was often the case. Due to the Su-9's limited range this was no easy task, as it was necessary to stage through airfields located no more than 1,000 km (620 miles) apart – something which was often impossible in Siberia and the Soviet Far East with their huge expanses of woodland. As a way out, the PVO top command proposed fitting the Su-9 with a special non-jettisonable large-capacity ferry tank instead of the standard drop tanks. In November 1959 Sukhoi OKB engineers proposed a conformal ventral tank fitted to the existing twin fuselage hardpoints in a manner similar to the *Dackelbauch* ('dachshund belly') ferry tank used on the Messerschmitt Bf 110. Estimated range with this tank was almost twice the range possible in standard configuration. Ironically, however, it was the military who eventually killed off the idea by demanding that a fuel jettison or emergency tank jettison feature be incorporated – which turned out not to be feasible.

Early-production Su-9s had a limited radius of action; depending on the flight profile, it was approximately 320-450 km (200-280 miles) when intercepting a typical target cruising at 20,000 m (65,600 ft). Later the

A typical publicity shot from the flight line of an IA PVO unit equipped with Su-9s. Typically of PVO aircraft, the tactical codes are blue. Note that the missile pylon leading edges have black anti-glare strips.

33

Above: Three Su-9s make a flypast in Vee formation during a flying display

Above: Technicians swarm like ants over a Su-9 undergoing maintenance

A crew chief places a boarding ladder against the side of a late-production Su-9 (c/n 1415343) loaded with a full complement of four RS-2-US missiles and drop tanks.

internal fuel capacity was increased, extending the radius of action to an acceptable 430-600 km (270-370 miles). The conversion training programme for Su-9 pilots included several typical operational scenarios, and the optimum climb profile and interception tactic were different in each case.

Intercepting a target at medium altitude presented no difficulties for the average pilot. An interesting 'pass me and die' tactic was developed against supersonic unmanned targets; the Su-9 would be directed to a point ahead of the target on its anticipated course, flying slightly below it, and then fire the missiles as the target overtook it. On the other hand, slow-flying high-altitude targets (such as the famous Lockheed U-2) were a much tougher nut to crack. In order to reach the Su-9's service ceiling the pilot needed to execute the so-called basic mode – ie, climb to a certain altitude (not less than 10,000 m/ 32,800 ft), then accelerate to Mach 1.6 in level flight and gently put the aircraft into a climb while maintaining an indicated airspeed of 1,100 km/h (683 mph), the Mach number increasing to 1.9 in the process. After that, the aircraft would climb at high angles of attack; the speed needed to be not less than Mach 1.7, otherwise the Su-9 would simply stall at that altitude. Concurrently the pilot had to make constant course corrections as instructed from the ground in order to get a target lock-on – and, since the interceptor was closing in on the target all too fast, the pilot had very little time for making an attack.

Slow-flying Yakovlev Yak-25RV high-altitude reconnaissance aircraft were used as practice targets for simulated attacks, as were Tupolev Tu-16 medium bombers and other Su-9s. During live weapons training at target ranges the Su-9s fired at Ilyushin IL-28M, Mikoyan/Gurevich M-15 and M-17 target aircraft (converted time-expired IL-28 tactical bombers and MiG-15/MiG-17 fighters respectively) and purpose-built Lavochkin La-17 jet-powered target drones.

Conversion training was something of a problem initially due to the lack of a dedicated trainer version. Future Su-9 pilots first made several flights in a subsonic UTI-MiG-15 trainer; then, having got a credit in the theoretical training course, they started mastering the practice part of the course on the single-seat Su-9. This situation persisted until 1962 when MMZ No.30 'Znamya Trooda' at Moscow-Khodynka started turning out dual-control Su-9U trainers. Several of these aircraft were delivered to the 148th TsBP i PLS, the first-line units equipped with the Su-9 also taking delivery of one trainer each. Even so, the UTI-MiG-15 (supplanted in the second half of the 1960s by the Su-7U) was still used for Su-9 pilot lead-in training because the Su-9Us were in short supply.

The Su-9 was famed for its 'willingness to fly', and the type's service record includes a few truly amazing incidents. For instance, on 11th June 1964 the crew of a 179th IAP (*istrebitel'nyy aviapolk* – fighter regiment) Su-9U, trainee Capt. Mel'nikov and instructor Maj. Nikolayev, lost concentration during the approach to Stryy AB after a training sortie, allowing speed to bleed off dangerously, and ejected, fearing that the trainer would stall and spin into the ground. Left to its own devices, the aircraft unexpectedly righted itself, climbed to about 1,300 m (4,265 ft) and circled around the airbase until it ran out of fuel. It then glided down and landed on its own (!) in a ploughed field; unfortunately the landing was less than perfect and the aircraft sustained major structural damage, being declared a write-off.

Another case when the aircraft displayed more presence of mind than the driver occurred just seven months later. On 25th January 1965 Lt. Col. Ovcharov took off from Sary-Shagan AB on a night training mission in a single-seat Su-9; soon afterwards he discovered a control system malfunction and promptly ejected. Came dawn next day, and the aircraft was discovered 32 km (19¾ miles) from the base in virtually undamaged condition, save for a punctured No.1 fuselage fuel tank! The aircraft was standing 'in the middle of nowhere' in flat steppeland, resting on its drop tanks which had been crushed by the impact. The investigation showed that the fuel had not even been used up completely, and it was sheer luck that there had been no fire. The pilotless Su-9 had touched down in a wings-level attitude at about 400 km/h (248 mph), the crushed drop tanks turning into improvised skis on which it slithered for about 250 m (820 ft) before coming to a standstill. The bottom line: this unique episode was classed as a 'non-fatal accident/aircraft repairable' and the Su-9 was actually repaired and returned to service.

Nevertheless the first months of operational service revealed a host of problems and shortcomings. For one thing, early-production AL-7F-1 engines had an appallingly low service life (a mere 25 to 50 hours), which meant that the interceptors were often grounded by the lack of engines. For another, systems and equipment often failed both in flight and during ground checks, and many equipment items were not easily accessible, requiring the aircraft to be substantially 'undressed'. An incessant stream of claims kept pouring in at the factory and the OKB; the latter took corrective measures, and at times the factory was simply swamped in documents from the OKB requiring this or that modification to be made. This unfortunate situation was eventually resolved, the OKB and the Air Force agreeing that the manufacturer's teams would perform all necessary modifications *in situ*. One of the first important upgrades was the introduction of the ESUV-1 electrohydraulic system controlling the air intake shock cone and auxiliary blow-in doors. In the course of the year 1960 teams consisting of Sukhoi OKB and Novosibirsk aircraft factory personnel had modified more than 120 Su-9s in service.

On the other hand, jet fuel was abundant in those days, and in the days of the Su-9's service introduction period the PVO pilots often logged as many as 150-200 flying hours per year. The service pilots were happy with the fighter's performance and handling; the Su-9 had next to no unpleasant peculiarities in transonic flight mode. True enough, the Su-9's outstanding acceleration characteristics created a few problems in the early days; pilots recalled that the aircraft accelerated very quickly to its maximum landing gear transition speed during take-off and you had to be quick in getting the gear up. Other peculiarities which took some adjusting to included an unusually high angle of attack during climb and landing approach, rapid deceleration from high speeds when the engine was throttled back, and the high landing speed.

Technicians prepare to remove the ejection seat from a Su-9 by means of a small crane. Note the partially open airbrakes. An APA-5 power cart based on the widespread Ural-375D 6 x 6 army lorry provides electric power; unusually, this one has a cab with a collapsible canvas top, a rare version.

Above: A Su-9 caught by the camera seconds after lifting off on a practice sortie or a checkout flight, as indicated by the absence of missiles. The red star on the tail appears to have vanished altogether.

Above: A pair of Su-9s coded '40 Blue' and '90 Blue' prepares to take off on a night training sortie. Again, neither aircraft carries missile.

'Though rain may pour and blizzards blow, our vigil still we keep.' This view of a Su-9 on quick-reaction alert (with missiles) illustrates the compound curvature of the intake centrebody's surface.

Meanwhile, the world political situation did not look any too good; the Cold War was gaining momentum. In the late 1950s the US Central Intelligence Agency launched a large-scale reconnaissance operation against the Soviet Union, and the Lockheed U-2 strategic reconnaissance aircraft figured most prominently in this operation. The CIA acted on the assumption that the USSR lacked the technical means for intercepting aerial targets flying above 20,000 m (65,600 ft). At first spy missions involving incursions into Soviet airspace were flown from Pakistani territory, the U-2 flying along the Soviet border. Then, as the Americans grew confident they could do it with impunity, in early 1960 the CIA decided to penetrate deep into Soviet territory. The first such intrusion took place on 9th April when a U-2 originating from Peshawar, Pakistan (and flown by none other than Francis Gary Powers), reconnoitred several weapons test ranges in Turkmenia. A unit equipped with Su-9s was stationed in the area, and several of the interceptors scrambled to get at the intruder; however, not yet being fully familiar with their mounts and receiving faulty instructions from ground control, the Soviet pilots failed to locate the U-2.

The fate of the aircraft, which had only just been recommended for official service acceptance, now hung by a thread. A special commission arrived at the unit's base to investigate the incident and find the culprit. The commission included test pilots Vladimir S. Ilyushin (representing the Sukhoi OKB) and L. N. Fadeyev (representing GK NII VVS); the latter pilot performed a checkout flight, using the officially approved technique, and reached the required altitude of 20,000 m. Thus the Su-9 was exonerated completely; the PVO top command vented its wrath on the IA PVO's Chief of Combat Training Lt. Gen. Pogrebnyak who was removed from office. For the two weeks that followed, test pilots Vladimir S. Ilyushin, Gheorgiy T. Beregovoy, Nikolay I. Korovushkin and L. N. Fadeyev (who had just competed the State acceptance trials of the T-3-51 weapons system) maintained combat duty at the GK NII VVS facility in Akhtoobinsk, ready to stop a new incursion, should it occur. On 26th April a squadron of Su-9s flown by service pilots arrived in Akhtoobinsk, relieving the test pilots of this duty.

On 1st May 1960 F. G. Powers ran out of luck while flying another spy mission in U-2A 56-6689. This time his route took him across the entire territory of the Soviet Union from south to north; the objective was to reconnoitre sensitive military installations in the Urals region. PVO fighter regiments stationed along his route repeatedly but fruitlessly tried to intercept him; they were equipped with the MiG-19PM which, with its service ceiling of

Above: A production Su-9 equipped with two drop tanks seen during final approach. It was standard operational procedure to deploy the flaps for landing only (compare to the top photo on the opposite page showing that the flaps are left up for take-off).

16,600 m (54,460 ft), was no match for the U-2. The only aircraft that stood a chance of getting at the high-flying spyplane was a brand-new Su-9 which happened to be staging through Sverdlovsk-Kol'tsovo airport; however, the Su-9's armament consisted solely of air-to-air missiles, and of course none were carried on the delivery flight. In a gesture of despair, the local PVO commanders ordered ferry pilot I. Mentyukov to take off and ram the intruder, and take off he did but failed to locate the target. Shortly afterwards another PVO unit accomplished the objective: Powers' aircraft was destroyed by an S-75 Tunguska (NATO codename SA-2 *Guideline*) surface-to-air missile. However, the victory was marred by a tragic 'friendly fire' incident. A few minutes earlier two 356th IAP MiG-19PMs had scrambled to intercept the U-2; when the fighters showed up on the air defence radar screen a panicky PVO officer mistook them for more enemies and gave order to fire. One of the MiG-19s was shot down by a second missile, the pilot Lt. (sg) Sergey Safronov losing his life; he was later awarded the Order of the Red Banner posthumously.

Despite the unimpressive initial results (which played into the hands of the 'missile lobby' supported by the head of state Nikita S. Khrushchov), these events demonstrated the need for a high-altitude interceptor. Thus by the mid-1960s nearly 30 PVO fighter regiments had re-equipped with the Su-9. For instance, in the Moscow PVO District the type was flown by the 28th IAP (Krichev AB), the 415th IAP (Yaroslavl'-Toonoshna airport) and an unidentified regiment based in Rzhev; in the 8th PVO Army (the Ukraine) the Su-9 saw service with the 90th IAP at Chervonoglinskaya AB, the 179th IAP at Stryy AB, the 894th IAP at Ozyornoye AB near Zhitomir and the 136th IAP at Kirovskoye AB on the Crimea

Crews speed to a pair of Su-9s during a practice scramble

Top: '75 Blue', an early-production Su-9 (as revealed by the non-functional gun blast plates), taxies out for take-off.

Above: Maintenance in progress? No, these three Su-9s – again early-production machines featuring gun blast plates – serve as ground instructional airframes. The MiG-19S fighters in the background might have been taken for an older type being replaced by a newer one, but the presence of the Tu-128 heavy interceptor further beyond indicates that the picture was taken at the PVO technical school in Solntsevo. Note the open filler cap of the No.1 fuselage fuel tank just above the wing leading edge.

Left: A trio of Su-9s carrying missiles and drop tanks makes an airshow performance.

Peninsula. The 2nd PVO Army based most in Belorussia included the 61st IAP in Baranovichi and the 201st IAP at Machoolishchi AB near Minsk operating Su-9s. As the Vozdookh-1 GCI system gradually found its way to PVO command centres, the Su-9 units began mastering the ground controlled intercept technique; the pilot would fly the aircraft in flight director mode, continuously making course corrections calculated by the command centre and relayed to the aircraft by the Lazoor' data link system. The aircraft was thus guided towards the target along the optimum trajectory, which considerably increased the 'kill' probability.

Unfortunately, like any other type, the Su-9 had its share of fatal accidents. The first one occurred on 8th March 1960 when a 61st IAP aircraft flown by Lt. (sg) Morgoon suffered an engine failure on take-off and the altitude was too low for safe ejection. The highest accident attrition was in 1961-63, the years when the Su-9 was being introduced *en masse* by the PVO. The most widespread causes of the accidents were engine surge, engine fuel control unit failures, electrics failures and hydraulic control actuator failures. Often the accidents occurred due to improper operation or maintenance by the flight and ground crews. As the Su-9 matured and the PVO units grew more familiar with it, operational reliability and accident rates improved to quite acceptable levels by the late 1960s.

In the summer of 1964 the 393rd GvIAP based near Astrakhan' began transitioning from the Su-9 to the Su-11 as the first unit to operate the type. The transition to the more advanced interceptor went fairly smoothly, and after Stage A of the evaluation programme had been completed by the regiment, the entire production run of the Su-11 was delivered to first-line units in the first half of 1965. Apart from the 393rd GvIAP, these were two Moscow PVO District units – the 790th IAP at Khotilovo AB and the 191st IAP in Yefremov.

Compared to the predecessor, the Su-11 was much more 'user-friendly' and safer to operate; the aircraft as a whole and the engine in particular were much more reliable, and there were extremely few accidents caused by hardware failures. On one occasion in 1968 a 191st IAP Su-11 flown by Capt. Yamnikov ingested a bird; the damaged AL-7F-2 engine started vibrating violently but kept running, enabling the pilot to make a safe emergency landing.

Another advantage over the Su-9 was the Su-11's greater maximum interception altitude thanks to the more powerful Oryol radar and the R-8M missiles; these allowed the aircraft to attack targets flying far above its own flight level and even gave it a limited

Above: A technician inspects the drop tanks of a Su-11. Traces of the removed tactical code '35' can still be seen on the nose.

Above: Inspection of the port tailplane of a Su-11.

A technician installs the air intake cover on a Su-11. The cover is marked '35', although the aircraft is coded '36 Red'.

Above and left: Two operational Su-11s coded '68' and '69' use their airbrakes to shorten their landing run; the brake parachute was not always deployed because collecting and repacking it afterwards was additional work for the ground crews. The aircraft obviously belong the same unit.

Below: Partly covered by tarpaulins, a Su-11 code '34' awaits the next mission.

Next page: A view of the 'flight line' at the PVO's Junior Aviation Specialists School in Solntsevo with Su-11 '36 Red' foremost. MiG-19s, including a radar-equipped MiG-19P, and MiG-17s can be seen across the line.

'look-down/shoot-down' capability. Interestingly, the operational tactics developed for the Su-11 included even attacks against ground targets and surface ships. In reality, however, this was never done; the idea was not pursued further than a single experiment with firing RS-2-US AAMs at ground targets from a Su-9 in 1966. This programme involved both GK NII VVS test pilots and service pilots, including some from the 350th IAP based at Belaya AB.

Taking account of the experience gained in the Vietnam War and the Arab-Israeli wars, the Soviet PVO command gave due attention to close-in dogfight tactics – even though the Su-9 and Su-11 were ill-suited for this kind of combat. Some units (for example, the 136th IAP at Kirovskoye AB even staged group dogfights). Deployments to forward operating locations (FOLs) were also practiced; for example, in March 1966 a group of 20 14th PVO Army Su-9s deployed to an ad hoc ice airfield in Tiksi, Yakutia, the ground personnel and ordnance being delivered there by Antonov An-8 transports. In the 1970s the 8th PVO Army practiced Su-9 operations from short tactical strips.

The Su-9 and Su-11 saw a good deal of 'actual combat', intercepting real-life targets. Thus in the late 1960s a pair of Su-9s scrambled from Kyurdamir AB, Azerbaijan, to intercept a pair of Iranian Air Force fighters that had intruded into Soviet airspace. Receiving authorisation from ground control, the flight leader fired a missile at the Iranians but the radar 'mistook' the two intruders flying in close formation for a single large aircraft and placed the missile dead centre – ie, the missile passed between the two fighters without hitting either of them. Far more often, however, the Su-9/Su-11 pilots had to deal with drifting reconnaissance balloons; in fact, a special version of the R-8M AAM with a larger warhead was developed for destroying these targets, since a standard missile was usually not enough to destroy the large balloon. For instance, in 1969 Ye. N. Kravets, a 179th GvIAP Su-9 pilot, fired RS-2-US AAMs at a reconnaissance balloon but succeeded only in shooting off the lower half of the extremely long sensor pack dangling under the balloon, so another fighter from a different unit had to be called in to finish it off. (Confusingly, there were two 179th GvIAPs – one at Krasnovodsk, Turkmenia, and the other at Stryy AB.) In an uncanny replay of this incident, in 1976 the same pilot (now flying a Su-11) again failed to completely destroy a reconnaissance balloon even with a special 'balloon killer' missile, annihilating only the lower half of the sensor pack.

Curious incidents also occurred from time to time. In the early 1970s a 393rd GvIAP Su-11 standing on quick-reaction alert (and armed with live missiles, of course) was scrambled to intercept a Su-9 acting as a practice target. In the course of a poorly organised intercept the Su-11 pilot fired a missile – and shot down his comrade-in-arms (who fortunately managed to eject and survived). A similar 'friendly fire' incident took place in 1969 in the 179th IAP (the one at Stryy AB). Maj. Koorilin on quick-reaction alert (QRA) duty was tasked with performing a practice intercept of a Su-9 from the same regiment. The pilot successfully located the target, got a good lock-on and, after receiving authorisation from the ground control centre, fired all four missiles. Realising what was about to happen, the aghast pilot pushed the stick forward, putting the aircraft into a dive; the semi-active radar homing missiles followed the radar beam, diving away from the target and self-destructing.

In the 1970s the Su-9 was gradually phased out, yielding its place to the MiG-25P/PD/PDS and the Su-15; as the units re-equipped with the new types, the 'unwanted' Su-9s were transferred to other units. From 1976 onwards the Su-9s were progressively retired as time-expired; units operating the type re-equipped with the MiG-23M, subsequently followed by the MiG-23ML and the ultimate MiG-23P developed specially for the PVO. The surviving Su-9s and Su-11 were assembled at the storage depots in Rzhev and Kuibyshev, where most of them succumbed to the elements. A few aircraft were converted into target drones; others survived as ground instructional airframes.

A few aircraft have been preserved for posterity as museum exhibits. The Soviet Air Force Museum (now Central Russian Air Force Museum) in Monino near Moscow has Su-9 '68 Red' (now falsely painted up as '68 Blue'; c/n 0615308) and Su-11 '14 Red' (similarly masquerading as '14 Blue'; c/n 0115307). A Su-9 coded '05' is preserved at the PVO Museum in Rzhev; another example coded '07 Blue' (c/n 100000510) is preserved at the base museum of the 148th TsBP i PLS at Savostleyka. The open-air museum at Moscow-Khodynka had a Su-11 coded '10 Red' (c/n 0715348); unfortunately the museum is now closed and the aircraft is in danger of being scrapped. One Su-9 is preserved at its birthplace at the Novosibisk Aircraft Production Association named after Valeriy P. Chkalov (NAPO – Novosi**beer**skoye aviatsi**on**noye proiz**vod**stvennoye obyedi**nen**iye), as factory No.153 is now called. Others were preserved at pioneer camps (children's summer recreation camps).

Above: Curious cadets swarm around Su-11 '36 Red' during a training session at the technical school in Solntsevo; the aircraft is jacked up for a landing gear retraction demonstration. Interestingly, all the cadets are carrying gas masks in cloth bags.

Su-11 '14 Red' (c/n 0115307) in the open-air display at the Central Russian Air Force Museum in Monino.

Chapter 4

Experimental Models

P-1 experimental two-seat interceptor

The Sukhoi OKB undertook the first design studies on a new fighter (subsequently known in-house as *izdeliye* P) in late 1954 to meet an MAP request. Both single-seat and two-seat versions were considered at this stage, as were different armament fits (cannons only, missiles only or a combination of unguided rockets and K-5 air-to-air missiles) and different powerplant options (Lyul'ka AL-11, Klimov VK-9F, Kuznetsov P-2 and P-4 axial-flow afterburning turbojets). Later the choice was narrowed to two versions; designated P-1 and P-2, they were powered by a single Lyul'ka AL-9 and two Izotov (Klimov) VK-11s respectively. (Sergey P. Izotov had by then succeeded Vladimir Ya. Klimov as head of the Leningrad-based OKB-117.) On 19th January 1955 the Soviet Council of Ministers issued a directive ordering the development of these aircraft. Detailed specifications for the new fighters were developed and approved by all parties concerned in the course of February-March 1955, enabling the OKB to begin more detailed design work on the general arrangement and systems layout.

Early project studies envisaged a nose air intake of the kind used on the Su-7, Su-9 and Su-11; however, the need to incorporate a powerful fire control radar with a large antenna dish in the nose made this impossible. (The radar type changed, too, as the design work progressed; the Almaz-7 (Diamond-7), Uragan and Pantera radars were considered consecutively.) Hence the designers quickly opted for variable lateral air intakes; in the course of 1955 this intake design was studied in detail, allowing two advanced development projects to be prepared by the end of the year.

Both versions were two-seaters, the crew consisting of a pilot and a weapons systems operator (WSO) seated in tandem under a common canopy; both versions featured a conventional layout with mid-set delta wings and lateral air intakes of semi-circular cross-section with a movable centrebody in the shape of a half-cone, as on the Dassault Mirage series or the Lockheed F-104 Starfighter. Actually the intakes bore but a passing resemblance to those of the Mirage or Starfighter; the leading edges were sharply raked in plan view instead of being at right angles to the fuselage centreline, forming an elliptical contour, and the centrebodies were much more pointed. The nose was occupied by a Pantera or Uragan radar in a large conical radome.

The first version (the P-1) was to be powered by a single AL-9 axial-flow afterburning turbojet rated at 10,000 kgp (22,045 lbst) in full afterburner, the inlet ducts merging at the compressor face, while the P-2 had the two VK-11s installed side by side. Another difference was that, in addition to the two K-7 AAMs common to both versions, the P-1 was to be armed with up to 32 70-mm (2.75-in) ARS-70 Lastochka (Swallow) folding-fin aircraft rockets (FFARs) in special automatic launchers or up to 30 85-mm (3.34-in) TRS-85 spin-stabilised unguided rockets (***toor***boreak***tiv***nyy sna***ryad***) in launch tubes built into the wing roots. The unguided rockets were intended as a close-range weapon for use against large and sluggish aircraft such as heavy bombers, not for ground attack.

A mock-up review commission convened in late 1955 to assess the two project versions; the P-1 was accepted for full-scale development and the P-2 axed, whereupon the detail design stage duly began in the first half of 1956. At this stage the design underwent major changes to suit the chosen operational concept. First of all, the rocket launchers mentioned above were rejected in favour of a different arrangement. Fifty 57-mm (2.24-in) ARS-57 Skvorets FFARs were to be housed in launcher packs located around the circumference of the nose immediately aft of the radome in a similar manner to the Lockheed F-94C Starfire; they were to be fired in a salvo or a ripple after six doors hiding the launchers had been opened. The radar was enclosed in a hermetically sealed capsule to prevent it from being affected by the rockets' exhaust gases. The designers chose not to area-rule the centre fuselage in order to maximise the internal fuel capacity.

Detail design of the P-1 was completed in August 1956, and as soon as the final blueprints were delivered to OKB-51's experimental production facility the latter commenced

The 'Soviet Thunderchief'. At this angle the P-1 experimental aircraft bore a certain resemblance to the Republic F-105B but the wing design was altogether different. The unusual air intake design and the wing dogtooth are clearly visible; the rocket bay doors can be seen aft of the radome.

43

Three-quarters rear view of the P-1. The main landing gear design was similar to that of the MiG-21, the wheels rotating around the struts during retraction to lie vertically in the wheel wells.

prototype construction. By the end of the year it became clear that the Lyul'ka OKB was facing major development problems with the AL-9 and would be unable to deliver the engine on schedule, so as a stop-gap measure the Sukhoi OKB decided to install the proven, albeit less powerful, AL-7F for the initial flight tests. With this engine the P-1 was rolled out in May 1957; on 10th June the dismantled aircraft was delivered to the OKB's flight test facility in Zhukovskiy to begin ground checks. M. Goncharov was appointed project engineer.

Bearing no markings other than the national insignia, the P-1 performed its maiden flight on 12th July 1957 with OKB test pilot Nikolay Korovushkin at the controls. The manufacturer's flight tests continued until 22nd September 1958 and were performed by N. Korovushkin and Eduard V. Yelian. The intended AL-9 engine never arrived, and as the aircraft showed disappointing performance with the provisional powerplant it did not proceed beyond the initial flight test stage.

For a while OKB-51 persisted with the P-1 concept, trying to create a viable combat aircraft using the same layout. In particular, the possibility of installing a bigger engine (the Tumanskiy R15-300 afterburning turbojet or the Lyul'ka AL-11) and alternative weapon fits was explored. Pavel O. Sukhoi repeatedly tried, both single-handedly and jointly with the 'subcontractors' responsible for the radar and other equipment, to win Air Force support for the P-1, but their efforts were thwarted by a singular lack of interest on the part of the prospective customer. Soon the Sukhoi OKB launched a new major programme (the T-37 heavy fighter), and the P-1 was relegated to experimental aircraft status and finally abandoned altogether; the sole prototype was scrapped.

The following is a brief structural description of the P-1.

Type: Single-engined two-seat supersonic interceptor designed for day and night operation in visual meteorological conditions (VMC) and instrument meteorological conditions (IMC). The airframe is of all-metal construction.

Fuselage: Semi-monocoque stressed-skin structure of basically circular cross-section. Structurally the fuselage consists of three sections: forward (section F-1), centre (section F-2) and rear (section F-3), the latter being detachable for engine maintenance and removal.

The *forward fuselage* incorporates includes the conical radome and radar bay, the unguided rocket launcher bays, the pressurised cockpit and the nosewheel well located under it. The cockpit features tandem seating for the pilot and WSO under a common three-piece canopy featuring individual upward-hinged sections over the two seats; the canopy blends smoothly into the fuselage spine.

The cockpit section is flanked by the engine air intakes of semi-circular cross-section blending gradually into the centre portion of the fuselage. The raked supersonic air intakes feature sharp lips and movable shock cones on the inboard side. To prevent boundary layer ingestion the intakes are set apart from the fuselage, the extended inboard lips acting as boundary layer splitter plates.

The *centre fuselage* is slightly waisted in accordance with the area rule where the two inlet ducts merge at the entrance to the engine bay; it incorporates wing attachment fittings and accommodates integral fuel tanks. The *rear fuselage* accommodates the engine and incorporates the tail unit attachment fittings and three airbrakes, one of which is located ventrally and the other two laterally.

Wings: Cantilever mid-wing monoplane with delta wings swept back 60° at quarter-chord. The leading edge features a prominent dogtooth similar to the one on the PT-8 prototype; the root portions incorporate provisions for installing unguided rocket launcher tubes. Two hardpoints for missile pylons are provided. The wings have one-piece Fowler flaps, with ailerons outboard of these.

Tail unit: Conventional swept cantilever tail surfaces comprising a one-piece fin with inset rudder and mid-set slab stabilisers (stabilators).

Landing gear: Hydraulically retractable tricycle type, with single wheel on each unit; the nose unit retracts aft, the main units inward into the wing roots and fuselage. All three struts have levered suspension and oleo-pneumatic shock absorbers; the nose unit is equipped with a shimmy damper. Originally a 660 x 200 mm (26.0 x 7.87 in) nosewheel and 900 x 275 mm (35.4 x 10.82 in) mainwheels were envisaged; due to design changes, however, they were substituted with a 570 x 140 mm (22.4 x 5.5 in) K-283 non-braking nosewheel and 1,000 x 280 mm (39.37 x 11.0 in) KT-72 brake-equipped mainwheels.

The nosewheel well is closed by twin lateral doors and a forward door segment, the mainwheel wells by triple doors (one segment is hinged to the front spar, one to the root rib and a third segment attached to the oleo leg). All doors remain open when the gear is down.

Powerplant: One Lyul'ka AL-7F axial-flow afterburning turbojet rated at 6,850 kgp (15,100 lbst) at full military power and 8,950 kgp (19,730 lbst) in full afterburner. The fuel capacity is 5,450 litres (1,199 Imp gal) and the fuel load 4,470 kg (9,850 lb).

Control system: Powered controls with irreversible hydraulic actuators throughout; a BU-49 rudder actuator, a BU-51 tailplane actuator and a BU-52 aileron actuator are used. Control inputs are transmitted to the aileron and stabilator actuators by push-pull rods, control cranks and levers; a combined control linkage utilising both rods and cables is used in the rudder control circuit. An AP-28Zh-1B autopilot is provided; the tailplane control circuit also features an AP-28 autopilot.

Hydraulics: Three separate hydraulic systems, each with its own engine-driven pump. The primary system operates the landing gear, flaps, airbrakes and air intake shock cones; the two actuator supply systems power exclusively the aileron, rudder and tailplane actuators.

Avionics and equipment: The avionics suite includes a Pantera fire control radar, an ARK-54 ADF, an RV-U low-range radio altimeter, an MRP-56P marker beacon receiver, SOD-57M DME, a GIK-1 gyro-flux gate compass, an AGI-1 artificial horizon, an RSIU-4V two-way VHF communications radio, an SPU-2 intercom, a data link receiver making up part of the *Gorizont-1* (Horizon-1) GCI system, an SRZO-2M Kremniy-2M IFF transponder and a Sirena-2 radar warning receiver.

Armament: The principal armament consists of two K-7S air-to-air missiles carried on underwing pylons. Additionally, 50 57-mm ARS-57 FFARs are carried in buried launchers enclosed by six powered doors around the circumference of the fuselage nose.

P-2 two-seat interceptor project

The P-2 interceptor developed in parallel with the P-1 differed from the latter mainly in having a powerplant consisting of two Izotov (Klimov) VK-11 axial-flow afterburning turbojets rated at 11,250 kgp (24,800 lbst) in full afterburner. Each engine breathed through a separate air intake of identical design to those of the P-1; the centre fuselage was not area-ruled. The estimated fuel load amounted to 3,200 kg (7,050 lb).

Unlike the P-1, the P-2's principal armament consisted of two 30-mm (1.18 calibre) Nudelman/Rikhter NR-30 cannons installed in the wing roots, with 100 rounds per gun. In maximum take-off weight configuration up to 20 TRS-85 or up to 16 ARS-70 unguided rockets could be carried under the wings. Alternatively, two K-7 AAMs could be carried and guided to the target by the Pantera radar installed in the nose. Due to the higher take-off weight as compared to the single-engined version the P-2 was to feature larger and stronger 1,000 x 275 mm (39.37 x 10.8 in) mainwheels.

Work on the P-2 project was discontinued shortly after the in-house project review in late 1955.

T-37 experimental interceptor

Development of a new high-altitude fast interceptor which later received the manufacturer's designation T-37 began in early 1958. The Air Force's requirements to achieve a service ceiling of 27,000 m (88,580 ft) and a speed of 3,000 km/h (1,860 mph) at high altitude meant that the aircraft had to be designed around an all-new and extremely powerful engine. The engineers could choose between two engines – the AL-11 developed by Arkhip M. Lyul'ka's OKB and

Above: This head-on view illustrates the P-1's elliptical-section forward fuselage and the circular air intakes, as well as the drooped wing leading edge outboard of the dogtooth.

Rear view of the P-1

the R15-300 created by Sergey K. Tumanskiy's OKB-300. However, the former engine did not even exist in hardware form, whereas the R15-300 was in the middle of its flight test programme (it was developed for the Mikoyan Ye-150 experimental heavy interceptor family); hence it was this engine that was selected to power the T-37.

In keeping with the prevailing weapons systems design ideology of the time the aircraft was considered as an integral part of a fully automated aerial intercept system, being nothing more than an air-to-air missile launch platform. Therefore the T-37 was to be equipped with a whole range of automatic guidance (ie, GCI) systems, with far-reaching plans to integrate an automated target detection, tracking and attack system enabling all-aspect engagement in both pursuit and head-on mode. The pilot's functions were thereby reduced to monitoring the system and taking corrective action if anything went wrong.

Officially development of the T-3A-9 aerial intercept weapons system (regarded as a refined version of the existing system based on the T-3 interceptor) was initiated by a Council of Ministers directive issued on 4th

June 1958. This document required OKB-51 to develop the T-3A interceptor and the K-9-51 armament system comprising two K-9 air-to-air missiles and the associated control system. The missile itself was developed by OKB-2-155, the aircraft weapons branch of the Mikoyan OKB (where it was known as the R-38), while the guidance system was the responsibility of KB-1 headed by Konstantin Patrookhin, a division of the Ministry of Defence Industry (now known as NPO Almaz, 'Diamond' Scientific & Production Association). The missile was to be part of the Uragan-5B automated aerial intercept system and be carried by the Mikoyan Ye-150 and Sukhoi T-37 (the armament systems for these aircraft were designated K-9-155 and K-9-51 respectively in accordance with the two design bureaux' code numbers). The K-9 had semi-active radar homing and the target would be illuminated by a powerful fire control radar designated TsP and installed identically on both aircraft (in the shock cone of the nose air intake).

In addition to the T-37 aircraft with its missile armament and radar, the T-3A-9 weapons system comprised the Looch-1 (Beam-1) GCI system, the Barometr-L data link/navigation/instrument landing system and the SRZO-2M Kremniy-2M IFF.

Actual design work on what was now designated T-37 commenced in June 1958, with I. E. Zaslavskiy as chief project engineer. The advanced development project stage was completed in the early spring of 1959; the aircraft was designed to carry not only K-9 AAMs but also 57-mm ARS-57M FFARs or 212-mm (8.34-in) ARS-212 high-velocity aircraft rockets (HVARs). The envisaged maximum speed at an altitude of 25,000 m (82,000 ft) was 3,000 km/h.

The T-37 shared the general arrangement of the earlier T-3, featuring an axisymmetrical nose air intake with a massive shock cone which also housed the radar dish, delta wings with 60° leading-edge sweep, and tail surfaces swept back 55° at quarter-chord with all-movable tailplanes. Unlike the predecessors, however, the aircraft featured a monocoque fuselage structure with no longerons (although Sukhoi OKB engineer Eduard S. Samoylovich, who was on the T-37 design team, claims the fuselage was of the usual semi-monocoque type). The integral fuel tanks and the inlet ducts were one-piece welded structures made of AMTs aluminium alloy; a similar technology was used for manufacturing the rear fuselage which was made of titanium and heat-resistant steel.

In a departure from normal design practice, the multi-shock intake centrebody was fixed; airflow control was exercised by means of a translating outer ring fitted to the forward fuselage, plus auxiliary blow-in doors on the fuselage sides.

The ADP was duly reviewed and approved at the OKB in the spring of 1959; detail design continued until the early summer, whereupon prototype construction began. Then, when the aircraft was already substantially complete, GKAT unexpectedly ordered all work on the T-3A-9 weapons system (according to some sources, it had been redesignated T-9M by then) to be terminated and all hardware manufactured under this programme to be destroyed. The unfinished airframe was removed from the assembly jigs and broken up – a sad end of what might have been a remarkable aircraft.

The following is a brief structural description of the T-37.

Type: Single-engined single-seat supersonic heavy interceptor designed for day and night operation in VMC and IMC. The all-metal airframe structure is made mostly of AMTs aluminium alloy.

Fuselage: Welded stressed-skin structure of circular cross-section; maximum fuselage diameter 1.7 m (5 ft 7 in).

Structurally the fuselage consists of two sections: forward (section F-1) and rear (section F-2), the latter being detachable for engine maintenance and removal.

The *forward fuselage* incorporates an axisymmetrical circular air intake with a translating lip (a hydraulically powered annular forward section) and a fixed multi-shock dielectric centrebody (shock cone). The latter is attached to a vertical splitter which divides the air intake into two ducts flanking the cockpit, the fuselage fuel tanks and avionics bays. The radar is housed in a hermetically sealed capsule ahead of the cockpit. The cockpit canopy comprises a fixed windscreen and an aft-hinged rear portion blending into a shallow fuselage spine.

The *rear fuselage* is a one-piece structure making large-scale use of titanium alloys; it accommodates the engine with its extension jetpipe and afterburner. The rear fuselage incorporates four airbrakes in a cruciform arrangement and a ventrally located brake parachute container. Originally an annular ejector was to be fitted around the engine nozzle to enhance thrust; on the actual aircraft it was replaced by eight air scoops located around the rear fuselage circumference.

Wings: Cantilever mid-wing monoplane with delta wings. Leading-edge sweep 60°, anhedral 3° from roots, incidence 0°.

The wings are of single-spar stressed-skin construction with three transverse beams (auxiliary spars) which, together with the main spar, form four bays: the leading edge, forward bay, mainwheel well and centre bay. The forward and centre bays delimited by the main spar/No.1 auxiliary spar and the Nos 2 and 3 auxiliary spars respectively (plus the root rib) act as integral fuel tanks, with the mainwheel wells in between. The upper and lower wing skins consist of large duralumin panels with integral stiffeners. Additional structural elements are installed between the Nos 1 and 2 auxiliary spars, acting as attachment points for the main gear units and their retraction jacks.

The trailing edge is occupied by hydraulically actuated one-piece Fowler flaps and ailerons; both are made of duralumin. One missile pylon is installed under each wing.

Tail unit: Conventional tail surfaces; sweepback at quarter-chord 55°. The *vertical tail* comprises a one-piece fin and an inset rudder; the fin root houses the rudder's hydraulic actuator. The fin is a single-spar structure with a rear auxiliary spar (internal brace), front and rear false spars, and chemically milled duralumin skins with integral stiffeners. The glass-fibre tip fairing incorporates a wire mesh antenna for the communications radio. The mass-balanced rudder is a single-spar structure; it is carried on three brackets.

The cantilever *horizontal tail* mounted 140 mm (5½ in) below the fuselage waterline consists of slab stabilisers (stabilators) rotating on raked axles; anhedral 5°, incidence in neutral position –2°. Each stabilator is a single-spar spot-welded structure with front and rear false spars and chemically milled duralumin skins with integral stiffeners. The stabilators are differentially controlled by separate hydraulic actuators.

Landing gear: Hydraulically retractable tricycle type, with single wheel on each unit; the nose unit retracts aft to lie under the cockpit, the main units inward into the wing roots. All three landing gear struts have levered suspension and oleo-pneumatic shock absorbers.

The steerable nose unit is equipped with a 570 x 140 mm (22.4 x 5.5 in) K-283 non-braking wheel and a shimmy damper. The main units have 800 x 200 mm (31.5 x 7.87 in) KT-89 mainwheels equipped with disc brakes. The nosewheel well is closed by twin lateral doors and a forward door segment linked to the oleo strut; each main unit has two door segments hinged to the oleo leg.

Powerplant (provisional): One Tumanskiy R15-300 axial-flow afterburning turbojet rated at 6,840 kgp (15,080 lbst) dry or 10,150 kgp (22,380 lbst) in full afterburner.

The R15-300 is a single-shaft turbojet with a five-stage compressor, a cannular combustion chamber, a single-stage turbine and an afterburner with a three-position variable nozzle. The compressor has bleed valves at the third stage. Starting is by means of a 150-hp S3 jet fuel starter connected to the ventral accessory gearbox. An electronic engine control system is provided.

EPR at take-off rating 4.75; mass flow at take-off rating 144 kg/sec (317 lb/sec), maximum turbine temperature 1,230° K. SFC 2.45 kg/kgp·h (lb/lbst·h) in full afterburner and 1.12 kg/kgp·h in cruise mode. Length overall 6,650 mm (21 ft 9⅞ in), casing diameter 1,640 mm (5 ft 4½ in). Dry weight 2,590 kg (5,710 lb).

The engine was to breathe through a circular supersonic air intake with a shock cone, augmented by lateral auxiliary inlet doors. The engine was to be cooled by boundary layer bleed air, with cooling air intake scoops at frames 25 and 29.

Control system: Powered controls with irreversible hydraulic actuators throughout.

Fuel system: Internal fuel is carried in six integral tanks (Nos 1 and 2 fuselage tanks, plus four wing tanks) and one bag-type tank (the No.3 fuselage tank); the total capacity is 3,870 litres (851.4 Imp gal). The fuel tanks feature a combined pressurisation/vent system ensuring normal fuel consumption, plus a special tank allowing the engine to run under negative-G conditions. There are provisions for carrying a 930-litre (204.6 Imp gal) drop tank under the centre fuselage.

Armament: Two K-9 (R-38) SARH air-to-air missiles carried on underwing pylons. Alternatively, 57-mm ARS-57 FFARs in pods or ARS-212M HVARs on launch rails can be carried.

Avionics and equipment: The avionics suite includes a TsP-1 fire control radar, an RSBN-2 Svod (Dome) short-range radio navigation (SHORAN) system with flush antennas built into the fin, an RV-U low-range radio altimeter, an MRP-56P marker beacon receiver with a flush antenna built into the fin, SOD-57M DME, a KSI compass system, a Put' (Way) navigation system, an RSIU-5V two-way VHF communications radio with a mesh-type antenna in the fin cap, a Lazoor' (Prussian Blue) data link receiver working with the Looch-1 or Vozdukh-1 (Air-1) GCI system, an SRZO-2M Khrom-Nikel' IFF transponder and a Sirena-2 radar warning receiver.

P-37 interceptor project

An alternative version of the T-37 featuring lateral air intakes was developed under the designation P-37. Unfortunately no details are known of this version.

Above: An artist's impression of the T-37 heavy interceptor from the ADP documents.
Below: Another artist's impression of the T-37. Note the very long intake shock cone, the annular nozzle ejector and the stalky landing gear.

T-49 experimental interceptor

Not wishing to put up with the obvious deterioration of the aerodynamics caused by the large radomes of the PT-8 and T-47, the Sukhoi OKB started investigating alternative radome/air intake arrangements in 1958. This led to the development of a highly unconventional and interesting design with two narrow curved air intakes flanking a conical radome. This arrangement ensured better operating conditions for the radar which was now fixed, ie, unaffected by the movement of the air intake shock cone. In order to minimise aerodynamic losses it was decided to incorporate a so-called isoenthropic air intake with a specially profiled inlet duct throughout its length. This was expected to significantly increase the efficiency of the air intake and hence of the powerplant as a whole;

The rear fuselage of the T-37 prototype in the assembly jig at the Sukhoi OKB' experimental shop. The aircraft was destined never to be completed.

Above: Head-on view of the T-49 development aircraft, showing the lateral air intakes of unusual shape (so-called sector arc intakes).

the inlet duct effectively became an 'additional compressor stage', increasing the engine pressure ratio perceptibly. At the same time the unusual intakes left ample space in the fuselage nose for the radar dish.

Bearing the in-house designation T-49, the aircraft incorporating the new intake design was development in accordance with a GKAT order dated 6th August 1957. It was decided to build the prototype by converting the unfinished T-39 airframe; the T-39 had been abandoned by 1958 because the latter aircraft had been demoted to experimental status, the programme under which it had been created becoming the responsibility of the Central Aero Engine Institute .

The conversion involved replacing the entire forward fuselage with a new assembly specially designed and manufactured at the OKB's experimental shop; the No.3 fuel tank, which had given place to a water ballast tank on the T-39, was reinstated in the process. The rebuilt aircraft was rolled out in October 1959 and M. S. Goncharov was put in charge of the test programme. After a period of ground checks OKB-51 test pilot A. A. Koznov successfully performed the T-49's maiden flight in January 1960. He reported that the aircraft handled well, displaying excellent performance; in particular, acceleration had improved considerably, just as the designers had expected.

Unfortunately the T-49's career turned out to be brief. In April 1960 the aircraft was damaged in an accident at Zhukovskiy; after this the T-49 underwent lengthy repairs and modifications but was destined to fly no more.

T-59 interceptor project

Another project which owed its existence to the T-37 heavy interceptor programme was the T-59. Like the P-37 described above, it featured lateral air intakes and was intended as a testbed for the TsP fire control radar. Eventually, however, the T-59 was never built and no other details are known.

Top and above: These two views illustrate the shape of the T-49's air intakes in side view; note the auxiliary blow-in doors immediately ahead of the cockpit windscreen. The aircraft type is marked on the tail. The would-be radome is of all-metal construction, since the aircraft was an aerodynamics research vehicle.

Chapter 5
Sukhoi Strikes Back

T-58 (Su-11M) single-engined interceptor (project)

In the second half of the 1950s the Western world began fielding new airborne strike weapons systems, forcing the Soviet Union to take countermeasures. Apart from missile systems, for which the Soviet government had a soft spot during the years when Nikita S. Khrushchov was head of state, new state-of-the-art interceptors possessing longer range and head-on engagement capability were required for defending the nation's aerial frontiers. Creating such an aircraft appeared a pretty nebulous perspective, considering that many a promising programme for the re-equipment of the VVS and the PVO was terminated during the Khrushchov era (including 35 aircraft projects and 21 engine programmes in 1958-59 alone).

In this generally troubled climate the outlook for the Sukhoi OKB seemed quite favourable at first. The Su-7 tactical fighter had been developed and put into production, with the Su-7B fighter-bomber version following hot on its heels; the Su-9 interceptor forming the core of the Su-9-51 aerial intercept weapons system had just sailed through State acceptance trials and entered for production at two major factories, achieving IOC with the PVO's fighter element. Concurrently the more advanced Su-11 derivative forming the core of the T-3-8M weapons system was undergoing trials, and the OKB was hard at work on the new T-3A-9 aerial intercept weapons system.

In February 1960, however, the clouds started gathering. OKB-51 was ordered to stop all development work on the T-37 heavy interceptor and the armament system created for it. By mid-1961 it became obvious that the first-line units of the VVS and the PVO had run into big problems with the recently introduced Su-7B and Su-9, the appallingly low reliability of the AL-7F engine being one of the worst. In the first 18 months of service, more than 20 aircraft were lost in accidents, more than half of which were caused by failures of the aircraft's single engine. The military demanded vociferously that the faults which came to the fore during the two types' service introduction period be corrected.

This situation heightened the Air Defence Force's interest in twin-engined aircraft which offered higher reliability. Without waiting for the results of the State acceptance trials the PVO began lobbying for the production entry of the Yak-28P interceptor on the grounds that it was twin-engined, ergo safer. The Ministry of Aircraft Industry amended its production plans accordingly, and in the three years to follow MAP's aircraft factories were to manufacture only twin-engined interceptor types – the Yak-28P and the Tupolev Tu-28 (Tu-128).

Additionally, the rival OKB-155 had begun State acceptance trials of the promising MiG-21PF light tactical fighter/interceptor; this aircraft was powered by a single Tumanskiy R11F2-300 afterburning turbojet which, while admittedly offering less thrust than the AL-7F, was much more reliable. Consequently on 27th November 1961 the Council of Ministers issued a directive ordering Su-9 production to be terminated in 1962 and cutting the Su-11's production run dramatically for the benefit of the Yak-28P which was to be produced at the same plant No.153 in Novosibirsk.

Thus OKB-51 was now facing not just further cuts in its programmes but the daunting prospect of being closed altogether for a second time as unnecessary. Considering the attitude of the nation's political leaders towards manned combat aircraft, the chances of developing all-new aircraft were close to zero; all the OKB could do was to modernise existing designs, and then only if state-of-the-art missile armament was integrated did these plans have any chance of success.

This was the situation in which the OKB started development of a new single-engined interceptor at its own risk; the aircraft was known in-house as the T-58 – the first aircraft to have this designation. To win support at the top echelon and avoid possible repercussions the project was disguised as a 'further upgrade of the T-3-8M weapons system'. The aircraft was to feature a fire control radar with increased detection range and a wider field of view and be armed with longer-range and more lethal missiles.

KB-339 offered two alternative radars for the T-58 – the Oryol-2 developed under chief project engineer G. M. Koonyavskiy as an upgrade of the RP-11 Oryol fitted to the production Su-11 and the brand-new Vikhr' (Whirlwind) developed under chief project engineer F. F. Volkov, a scaled-down version of the Smerch (Tornado) fire control radar created for the Tu-28 long-range heavy interceptor. Both of these radars, however, were too bulky to fit inside the shock cone of an axisymmetrical air intake as used on the Su-11; the only option was to use the entire forward extremity of the fuselage for accommodating the radar and utilise lateral air intakes.

By then OKB-51 already had some experience with lateral air intakes, having used them on the P-1 and T-49 experimental interceptors; also, such intakes were envisaged for one of the T-37's PD project versions. As distinct from all these aircraft, the T-58 featured two-dimensional (rectangular-section) intakes with vertical airflow control ramps – a design that was not yet fully explored in the Soviet Union at the time. The fuselage design also drew on one of the T-37's PD project versions, featuring a complex shape; the perfectly conical radome mated with a basically cylindrical forward fuselage that was flattened from the sides in the cockpit area and flanked by the air intakes. The latter blended smoothly into a centre fuselage which was again of basically cylindrical shape; the fuselage structure aft of frame 18 was identical to that of the Su-11, as were the wings, tail unit, landing gear, control system and powerplant – one AL-7F-2 engine. To appease the nation's missile-minded leaders, two air-to-air missiles were envisaged as the T-58's sole armament.

Development work progressed quickly; as early as July 1960 the first metal was cut on the T-58 prototype and assembly of the forward fuselage began at MMZ No.51. But then the appetites of the military started growing, the customer demanding ever-higher performance. In particular, the new interceptor was required to have all-aspect engagement capability against targets flying at altitudes up to 27,000 m (88,580 ft) and speeds up to 2,500 km/h (1,550 mph) – ie, be capable of shooting them down in both pursuit and head-on mode.

In November 1960 the Council of Ministers drafted a directive ordering OKB-51 to equip the new interceptor with the Vikhr'-P fire control radar and the *Polyot* (Flight) automatic control system (sic – ie, GCI system). The armament was to comprise two K-40 AAMs; the choice of the missile type was

49

dictated by the military who also envisaged this weapon for the Mikoyan OKB's new interceptors. The aircraft was allocated the service designation Su-15 (Soviet fighters traditionally received odd-numbered designations but the designation Su-13, which should have followed the Su-11, was not used for reasons of superstition). The aerial intercept weapons system based on the future Su-15 received the provisional designation T-3-8M2 because the aircraft would be armed with K-8M2 missiles pending availability of the intended K-40 AAM; the K-8M2 was a refined version of the K-8M which, unlike its precursor, had all-aspect engagement capability.

At the end of 1960 the Sukhoi OKB began testing the T-58's powerplant on a special ground rig in order to verify the operation of the lateral air intakes. Meanwhile, construction of the flying prototype and a static test airframe proceeded at MMZ No.51. Still, time passed but the promised K-40 missile was nowhere in sight, and neither was the anticipated CofM directive concerning development of the Su-15 with the Vikhr'-P radar (exactly for this reason). Hence OKB-51 continued development of the interceptor with the alternative Oryol-2 radar and K-8M2 AAMs (the latter were a product of Matus R. Bisnovat's OKB-4); in the project documents this version of the aircraft was called Su-11M (*modifitseerovannyy* – modified, or *modernizeerovannyy* – upgraded). The flying prototype was due for completion in September 1961, but the work on the T-58 project was suspended in the summer of that year. 'For want of a missile the fighter was lost'? Well, not exactly; see the next paragraph.

T-58 (T-58D) twin-engined interceptor (project stage)

As an insurance policy in case the single-engined T-58 was rejected, in late 1960 the Sukhoi OKB prepared a new version of the project envisaging installation of two R21F-300 axial-flow afterburning turbojets side by side in the rear fuselage. This advanced engine rated at 7,200 kgp (15,870 lbst) in full afterburner was developed by OKB-300's new Chief Designer N. G. Metskhvarishvili. At the insistence of the PVO's General Headquarters the twin-engined version – likewise known officially as the Su-15 – was to be equipped with the Vikhr'-P radar and armed with two K-40 AAMs. The customer was adamant in this issue, even though GKAT pointed out that making use of the Oryol-2 radar and K-8M2 AAMs would allow the new interceptor to enter service much sooner. The PVO also specified that the aircraft should have the proposed general arrangement with lateral intakes and a pronounced 'waist' at the wing/fuselage joint in accordance with the area rule.

The general arrangement and internal layout of the T-58 was finalised in 1961. The powerplant was changed at this stage. On the one hand, the single-engined version was indeed rejected, the customer expressly demanding higher reliability which only a twin-engined aircraft could provide. On the other hand, the R21F-300 turned out to have serious design flaws, and after an uncontained engine failure which led to the loss of the Mikoyan Ye-8/1 development aircraft on 11th September 1962 further development of this engine was terminated. These circumstances prompted the Sukhoi OKB to select the proven R11F2-300 (aka *izdeliye* 37F2) to power the T-58. Accommodating the two R11F2-300s in the rear fuselage presented no problem; OKB-51 already had some experience with a similar engine installation on the T-5 development aircraft (see Chapter 1).

In the course of 1961 the OKB completed the detail design of the twin-engined interceptor whose in-house designation was now amended to T-58D. To this day it is not clear whet the D suffix stands for; it could mean both ***dvig**ateli* (engines), referring to the fact that there were now two engines instead of one, or *dora**bot**annyy* (modified or improved). Construction of the prototype and the static test airframe continued in the meantime – the unfinished airframes of the cancelled single-engined version were modified right in the assembly jigs. Another important change occurred at this point; since the K-40 AAM had been selected as the main armament for OKB-155's new Ye-155P heavy interceptor (the future MiG-25P), it was agreed that the T-58D (aka Su-15) would be armed with upgraded K-8M2 missiles, provided that a further improved version of the Oryol-2 radar was used. This version featuring a 950-mm (37⅜ in) antenna dish received a new name, Sobol' (Sable).

The T-58D, which drew heavily on the detail design project of the original single-engined T-58, utilised the classic layout with mid-set and conventional tail surfaces featuring an all-movable horizontal tail. Interestingly, contrary to normal Soviet practice, no chief project engineer was assigned to the T-58D until the mid-1960s. Pavel O. Sukhoi resolved the key issues related to the interceptor's design, while the problems arising in the course of day-by-day work were handled by his deputy Yevgeniy A. Ivanov.

As already mentioned, to speed up the development process it was decided to borrow the Su-11's wings and tail unit for the T-58. The prototype and the static test airframe were built using stock Su-11 sub-assemblies which were suitably modified to match the different fuselage shape at the wing/fuselage joint. This, in turn, allowed the Su-11's main landing gear design to be retained as well. The same applied to the tail surfaces; thus the fuselage was the only major component designed from scratch.

In its prototype form the T-58D was one of the few Soviet aircraft to make use of the area rule. The fuselage was 'squeezed' in two areas – near the cockpit (ie, just ahead of the air intakes) and at the wing/fuselage joint where a prominent 'waist' existed. This design was meant to minimise interference drag at transonic speed. (It may be said in advance that this 'waist' was eliminated on production examples, though not for aerodynamic reasons.)

The twin-engine powerplant increased reliability significantly over the single-engined version planned originally. Quite apart from the obvious case of engine failure (the probability of both engines failing for unrelated reasons is virtually negligible), having two engines allowed hydraulic and electric systems to be duplicated, using hydraulic and electric power sources fed by different engines. Thus the T-58D's hydraulic system comprised four separate subsystems – two main circuits (Nos 1 and 2) and two control system circuits (port and starboard) serving solely the control surface actuators; each circuit featured its own hydraulic pump.

Since the new aircraft would obviously be a lot heavier than the precursor while the wing area remained unchanged, the T-58D's field performance would clearly deteriorate as compared to the Su-11. To compensate for the added weight the engineers decided to use blown flaps (then referred to in Soviet terminology as *reak**tiv**nyye zakryl**ki* – lit. 'jet flaps'), the air for these being bled from the engine compressors; hence the area-increasing Fowler flaps of the Su-11 were replaced with simple flaps. The landing gear could not be left unchanged after all – it had to be reinforced to cater for the higher weight.

The general belief was that the T-58D would have to deal primarily with single low-manoeuvrability targets flying at altitudes between 2,000 and 24,000 m (6,560-78,740 ft) and speeds up to 2,500 km/h (1,550 mph). Without having a significant advantage in speed the interceptor stood no chance of destroying such targets in pursuit mode; hence high-speed targets were to be intercepted in head-on mode, and both tactics would be used against slower aircraft. The technique of intercepting targets flying at altitudes beyond the fighter's service ceiling had been perfected with the Su-11. It involved climbing to a so-called base altitude where the fighter would be guided towards the target by GCI control, subsequently detecting it with its own radar and getting a lock-on; after coming within missile launch range the fighter would pull up, firing the missiles in a zoom climb. The minimum missile launch altitude was restricted by the imperfections of the radar which lacked 'look-down/shoot-down' capability.

The creators of the weapons system intended to maximise its capabilities by automating the intercept procedure insofar as possible. To this end the T-58D was to feature a purpose-built automatic flight control system (AFCS) including heading adjustment command modules and pre-programmed optimum climb profiles. In the course of GCI guidance and the actual intercept the pilot could choose between three control modes – manual, semi-automatic (flight director mode) and fully automatic. OKB-51 undertook to perform the main part of the design work on the AFCS; until the system was fully up to scratch the prototypes would be equipped with AP-28T-1 autopilots.

The obligatory in-house review of the ADP and the mock-up review commission were dispensed with, since the T-58D was considered to be nothing more than an upgraded version of the Su-11. Only the cockpit section received the attentions of a mock-up review commission.

As already mentioned in Chapter 2, on 5th February 1962 the Council of Ministers issued a directive clearing the T-3-8M (Su-11-8M) weapons system for service. One of the document's items read, *'For the purpose of enhancing the system's combat capabilities the Su-11 aircraft is to be upgraded in order to enable attacks in head-on and pursuit mode against targets flying at altitudes between 2,000 and 24,000 m and speeds up to 2,500 km/h, as well as to further increase the system's reliability, ECM resistance and automation levels.'* Thus the Su-15 took the first steps towards fully legal status.

All the while the OKB actively undertook research and development work under the T-58D programme, using various test rigs. Among other things, a series of wind tunnel tests was held at TsAGI; ground rigs were built for testing the T-58D's electric and hydraulic systems, and the static test airframe was successfully tested to destruction. The creation of the AFCS, which was subsequently designated SAU-58 (*sistema avtomaticheskovo oopravleniya* – automatic control system), was a major R&D effort in its own right; in the course of its development OKB-51 made its first large-scale use of mathematical analysis and simulation. As early as 1961 the OKB built a special simulator for verifying the SAU-58; some of the system's components were tested in flight on the T47-4 development aircraft (see Chapter 2).

T58D-1, T58D-2 and T58D-3 interceptor prototypes

Designated T58D-1, the first prototype of the new interceptor was built as an aerodynamics

Above: The forward fuselage of the T58D-1 prototype during construction at the Sukhoi OKB's experimental shop at Moscow-Khodynka. This view illustrates the design of the port air intake.

The T58D-1 takes shape at MMZ No.51. The pure delta wings, the fairing between the engine nozzles and the slab stabilisers are clearly visible.

test vehicle; hence no radar was fitted, its place being occupied by test equipment. The aircraft was intended for stability/handling trials and performance testing to determine the maximum speed, range, service ceiling, fuel consumption rates and acceleration characteristics with and without external stores.

Upon completion in the first quarter of 1962 the aircraft was trucked to the LII airfield in Zhukovskiy; R. Yarmarkov was appointed engineer in charge of the test programme. On 30th May 1962, having duly passed the prescribed ground systems checks and taxying tests, the T58D-1 successfully performed its maiden flight with Sukhoi OKB chief test pilot Vladimir S. Ilyushin at the controls. Originally the aircraft wore the very appropriate but equally non-standard serial '58-1' which was later changed to '31 Blue'.

By the end of the year the first prototype had made 56 flights under the manufacturer's flight test programme and displayed good performance and handling, largely meeting the expectations of its creators.

Even at this stage of the tests the customer demanded installation of a new fire control radar. Accordingly the OKB undertook a redesign of the second and third prototypes' extreme nose and rearranged the cockpit instrumentation on these aircraft. On 17th September 1962 GKAT issued an order specifying that the T-58D should be equipped with a Smerch-AS fire control radar and armed with two K-8M2 missiles, one featuring semi-active radar homing and the other infrared homing. Thus all three advanced interceptors then under development – the Ye-155P, Tu-128 and T-58D – would feature the same radar type.

However, a change of radar would necessitate a redesign of the T-58D's forward fuselage; also, no prototype radar was available for installation. Meanwhile, early flight tests had shown that the new Sukhoi interceptor

Above: Appropriately serialled '58-1' and marked T-58 on the tail, the first prototype is seen here during manufacturer's tests. The short radome and the the short tail are readily apparent. The aircraft carries two drop tanks, a K-98R missile under the port wing and a K-98T (c/n 0205) under the starboard wing.

flew well and could be rapidly put into production; the integration of a new weapons control system with all the accompanying test and debugging work could delay production and service entry for years. The implications were obvious: the Sukhoi OKB had to push for a decision to use the upgraded Oryol radar on the T-58D – initially at least – at all costs.

Pavel O. Sukhoi was supported in this issue by Matus R. Bisnovat; together they succeeded in making their point not only to GKAT's top executives but also to the Commanders-in-Chief of the VVS and the PVO. On 2nd March 1963 Sukhoi and Bisnovat wrote to Council of Ministers Vice-Chairman Dmitriy F. Ustinov, explaining the situation. On 13th March Ustinov wrote back, stating his agreement to use the modified Oryol radar during the first stage of the trials and possibly the initial production stage – with the proviso that the Smerch radar would be integrated eventually. At the same time the military agreed to curb their appetites a little, reducing the target's maximum speed to 2,000 km/h (1,240 mph) and the maximum interception altitude to 23,000 m (75,460 ft). The State acceptance trials of the T-58D were slated for completion in early November 1963.

Thus the Sukhoi OKB managed to buy some time. The new Sobol' radar specified originally was never installed; instead the second and third prototypes (the T58D-2 and T58D-3) were to feature the Oryol-D radar, the D standing for *dorabotannyy* (modified). Changes made at this stage included relocation of the brake parachute container from the rear fuselage underside to the base of the rudder and provision of a ventral fin that folded sideways during landing gear extension to provide adequate ground clearance (a feature which later appeared on the MiG-23 tactical fighter). In fact the ventral fin was never to be installed, as the increased-area fin (see below) ensured adequate directional stability. The landing gear featured new wheels – the nose unit had the 600 x 155 mm (23.6 x 6.1 in) KT-104 wheel replaced with a larger 660 x 200 mm (26.0 x 7.87 in) KT-61/3 wheel, while the 880 x 230 mm (34.6 x 9.0 in) KT-69/4 mainwheels of the first prototype gave place to identically sized KT-117 wheels featuring more effective brakes cooled by a water/alcohol mixture.

In early 1963 the T58D-1 underwent modifications to the same standard as the second and third prototypes. At a very early stage of the tests Ilyushin had noted that the T-58D had poor directional stability as compared with the Su-11. The problem was cured by inserting a 400-mm (15¾ in) 'plug' at the base of the fin; this additional section conveniently provided accommodation for the brake parachute container.

Until the end of the year OKB test pilots Vladimir S. Ilyushin, Yevgeniy S. Solov'yov and A. T. Borovkov made a total of 104 flights in the modified T58D-1. New changes were progressively introduced into the design. In particular, to reduce drag the original radome was replaced with a more pointed one featuring a cone angle of 20° instead of 32°. The shape of the 'pen nib' fairing between the engine nozzles was also optimised to cut drag and the possibility of further reducing drag by forcing air through the engine bays was explored.

Due to the delayed decision on the model of radar to be used the second prototype (T58D-2) did not begin its test programme until April 1963; coded '32 Blue', it first flew on 4th May at the hands of Vladimir S. Ilyushin, with V. Torchinskiy as engineer in charge. Outwardly it differed from the first prototype in having a longer and more pointed radome with a cone angle of 28° but the vertical tail was still unmodified and the brake parachute container was accordingly located ventrally. The T58D-2 featured a complete avionics fit, including an Oryol-D58 radar (*izdeliye* 303D).

Tests of the avionics suite on the second prototype as part of the manufacturer's flight test programme continued until the end of June; in early August the T58D-2 was submitted for State acceptance trials which began in Zhukovskiy on 5th August 1963. Since development of the SAU-58 AFCS was running late, it was decided to hold a separate trials programme when the system became available. A State commission chaired by IA PVO C-in-C Air Marshal Yevgeniy Ya. Savitskiy was appointed for holding the trials of the AFCS. Interestingly, in official Ministry of Defence documents the T-58D was still referred to as the Su-11M, not as the Su-15.

The main part of the second prototype's State acceptance trials at LII was performed by GK NII VVS project test pilots S. A. Lavrent'yev, L. N. Peterin and V. I. Petrov plus Sukhoi OKB chief test pilot Vladimir S. Ilyushin. Other pilots who flew the aircraft at this stage were test pilots A. A. Manucharov, P. F. Kabrelev, V. G. Ivanov, I. I. Lesnikov and E. N. Knyazev, as well as State commission chairman Air Marshal Savitskiy and GK NII VVS Vice-Director A. P. Molotkov. Engineer Zhebokritskiy supervised the ground support and maintenance team, while engineer Lozovoy was in charge of the trials programme.

To speed up the process the State acceptance trials were held in a single stage instead

The third prototype (T58D-3) was coded '33 Blue'. The longer, more pointed radome and the taller vertical tail with a radar warning receiver antenna at the top are obvious in this view. Note the non-standard pylon-mounted antenna array under the nose and the five stars under the cockpit marking missile launches.

of the usual two. The trials programme included a series of spinning tests; this part of the programme involving seven flights was performed by LII test pilot Oleg V. Goodkov in September 1963.

On 2nd October 1963 OKB-51 test pilot Ye. K. Kukushev took the third prototype (T58D-3, '33 Blue') up for its maiden flight; A. Sholosh was engineer in charge of this aircraft's manufacturer's tests. The T58D-3 differed from the two preceding prototypes in having the internal fuel capacity increased by 180 litres (39.6 Imp gal) and a new AP-46 autopilot fitted.

The Air Force test pilots were generally pleased with the aircraft's handling but pointed out that aileron authority decreased at low speeds, complicating landings in a crosswind. Another criticism was that the engines tended to run roughly during certain vigorous manoeuvres with a sideslip. The take-off and landing speeds were rather high, too. These shortcomings resulted from the aircraft's chosen layout; thus, the unstable engine operation during manoeuvres with a sideslip was caused by the lateral air intakes – the intake on the opposite side to the direction of the sideslip was blanked off by the fuselage. The high take-off and landing speeds and unimpressive field performance had a different cause. As already mentioned, the T-58D featured blown flaps but the boundary layer control (BLC) system was inactive during the State acceptance trials because the version of the R11F2-300 engine featuring bleed valves for the BLC system was still unavailable.

The acceleration parameters had deteriorated as compared to the Su-11. On the other hand, no criticism was given of the stability and handling, except for a bit of instability in landing configuration at 340-450 km/h (211-279 mph). Also, remember that the twin-engined T-58D was heavier than the single-engined Su-11; not only the all-up weight but also the empty weight was 1.5 tons (3,306 lb) higher – 10,060 kg (22,180 lb) versus 8,560 kg (18,870 lb).

On 8th and 11th October 1963 respectively the second and third prototypes were flown to the GK NII VVS facility at Vladimirovka AB to continue the State acceptance trials – specifically, to perform live missile launches against real targets. This stage lasted from August 1963 to June 1964. The radar's performance was assessed for starters, using real aircraft as 'targets'; these included Tu-16 and IL-28 bombers, Yak-25RV high-altitude reconnaissance aircraft and the special Su-9L development aircraft equipped with an angle reflector to increase the RCS.

Actually the State acceptance trials involved not just the aircraft but the entire aerial intercept weapons system built around it.

These three views of the T58D-3 at a later date (note the different style and position of the tactical code) show the brake parachute container at the base of the fin and the rear aerial of the Pion antenna/feeder system above it requiring the rudder to be cut away from below. The undernose antenna pod is gone.

Designated Su-15-98, the system comprised the Su-15 interceptor powered by two R11F2-300 engines, its Oryol-D58 fire control radar (weapons control system) and two modernised K-8M1P air-to-air missiles which received the new designation K-98, alias *izdeliye* 56. The missile came in two versions (IR-homing and radar-homing). The system worked with the Vozdookh-1M GCI system.

The trials of the Su-15-98 weapons system progressed smoothly like never before, the customer expressing almost no criticisms. In December 1963 the chairman of the State commission, Air Marshal Savitskiy, endorsed a protocol formulating the preliminary results of the trials. This document stated that by early December a total of 87 flights had been made under the State acceptance trials programme, including 53 flights accepted 'for the record', 13 training flights, 16 flights 'off the record' and five positioning and checkout flights. The aircraft's flight performance and the operation of its systems and equipment had been fully explored; the performance figures obtained during the trials largely matching the manufacturer's estimates. In the course of the manufacturer's flight tests and State acceptance trials the three prototypes had made more than 300 flights without malfunctions, demonstrating the high reliability of the aircraft as a whole and its systems.

Being a highly experienced combat pilot who had seen a lot of action in the Great Patriotic War (his 'private' callsign *Drakon*, 'Dragon', dated back to those days), and a representative of the aircraft's customer into the bargain, Savitskiy would be ill advised to squander praise on aircraft which were not worthy of it. Thus his positive appraisal of the T-58D testifies to the undoubted success achieved by the Sukhoi OKB. Later, Sukhoi CTP Vladimir S. Ilyushin gave a similar account of those days, noting that the T-58's trials programme had been remarkably rapid and trouble-free.

The first K-8M1P (K-98) missiles were delivered to Vladimirovka AB in early 1964, allowing the T-58D's armament trials to begin; these included missile attacks in head-on mode. The missile basically met its specifications, except for the inability to score guaranteed 'kills' against high-speed targets because the proximity fuse could not detonate the warhead in time at high closing speeds. The verdict was that in a head-on attack the interceptor's missile armament enabled it to destroy targets doing up to 1,200 km/h (745 mph). The maximum interception range (combat radius) and effective range turned out to be shorter than expected, ferry range at optimum altitude with two drop tanks being only 1,260 km (780 miles) instead of the required 2,100 km (1,300 miles). Hence the State commission recommended that the internal fuel supply be increased even before the Smerch-AS radar envisaged for the second stage of the development programme was integrated.

The Sukhoi OKB decided to increase the fuel capacity by eliminating the 'waist' of the area-ruled fuselage. Within a short time the T58D-1 was modified in the first quarter of 1964 by riveting on a sort of 'corset' over the narrow portion of the centre fuselage so that the latter now had constant width determined by the distance between the outer faces of the air intakes. This increased the internal volume sufficiently to provide an internal fuel capacity of 6,860 litres (1,509 Imp gal) or 5,600 kg (12,345 lb) – more than the total pre-modification fuel capacity with drop tanks. Additionally, to improve stability and handling the aileron travel was increased from 15° to 18°30' and the air intake ramp adjustment time was reduced from 12 to 5-6 seconds.

On 2nd-16th June 1964 the converted T58D-1 underwent a special test programme at GK NII VVS to check the effect of the modifications; these received a positive appraisal and were recommended for introduction. The T58D-3 was later modified in the same fashion. On 19th June Yevgeniy Ya. Savitskiy made two flights in the T58D-2, and the State acceptance trials were officially completed on 25th June. By then the three prototypes had made a total of 250 flights between them, including 194 flights under the State acceptance trials programme (146 flights for the record and 30 training, positioning and checkout flights) and 26 demonstration flights. The trials included 45 missile launches and two emergency ordnance drops; nine target drones (seven IL-28Ms, one Yak-25RV-II and one MiG-15M) were destroyed.

The final report of the State acceptance trials (which referred to the weapons system as the 'Su-11-8M – Stage One upgrade') said that the new aircraft offered significant advantages over the production Su-11, especially as far as head-on engagement capability was concerned. Other strong points included greater flight safety, longer target detection/ lock-on range, a lower minimum operational altitude and the new radar's better ECM resistance. The report went on to say that the T-58 could intercept targets at medium and high altitudes round the clock and in adverse weather, the aircraft was easy to fly and could be mastered by the average pilot; the aircraft basically met the Air Force's operational requirements and was recommended for service entry. (The military pilots who flew the T-58, however, were somewhat less enthusiastic about the aircraft.) The upgraded T58D-3 was recommended by the State commission as the *etalon* (standard-setter) for production. It was recommended to explore the possibility of operating the Su-15 from unpaved runways after the wheels had been replaced with skids.

In the years that followed, all three prototypes were used as 'dogships' for testing new systems and equipment. Among other things, OKB-51 undertook a major programme aimed at refining the new interceptor's powerplant jointly with OKB-300, LII and GK NII VVS. Under this programme the T58D-1 and T58D-2 were used in 1964-65 to verify a new air intake control system intended to lift certain restrictions on powerplant operation when the engines were throttled back. The work went on and on but the efforts were fruitless and the restrictions stayed in force throughout the Su-15's service career.

In early 1965 the first prototype was modified for testing new cranked delta wings of increased area and 720 mm (2 ft 4⅜ in) longer span, with the leading-edge sweep on the outer portions reduced from 60° to 45°. The purpose of this modification was to increase aileron efficiency at low speeds. Vladimir S. Ilyushin made the first post-modification flight on 22nd February; by June three versions of the modified wings had been tried, the engineers settling for a version on which the outer wing leading edge was extended forward and cambered 7°. The new wings had a positive effect on both stability/handling and field performance and were subsequently incorporated on the Su-15TM (which see).

Later the T58D-1 was modified even more extensively to become the T-58VD short take-off and landing (STOL) technology demonstrator (also described in detail below). The second prototype was also modified for test work, becoming the T-58L (or T58-L) development aircraft. The T58D-3 was used to investigate a new technique involving the use of flaps for take-off (until then, the normal procedure was to deploy the flaps for landing only). The new flap settings were 15° for take-off and 25° for landing; this technique reduced the unstick speed by 20 km/h (12.5 mph) and the take-off run by an average 150 m (490 ft), and the modification was recommended for production. Boundary layer control was necessary to reduce the approach/ landing speed, but the modified engines featuring BLC bleed valves were still unavailable. The T58D-3 was also used for perfecting the SAU-58 AFCS and the upgraded Oryol-D58M radar in 1965-68.

The following brief structural description applies to the pre-modification T58D-1.

Type: Twin-engined single-seat supersonic interceptor designed for day and night operation in VMC and IMC.

Fuselage: Semi-monocoque riveted stressed-skin structure. The forward fuselage has a circular cross-section immediately aft of the conical dielectric radome; it features large detachable panels on the sides for access to the forward avionics bay. The latter is followed by the cockpit equipped with a KS-3 ejection seat and enclosed by a sliding bubble canopy similar to that of the Su-9/Su-11. A second avionics bay is located aft of the cockpit.

The forward fuselage is flanked by two-dimensional supersonic air intakes with sharp lips. Each intake features a two-shock vertical airflow control ramp with a movable panel governed by the UVD-58D engine/intake control system, plus a rectangular auxiliary blow-in door on the outer face. To prevent boundary layer ingestion the intakes are set apart from the fuselage, the inner lip acting as a boundary layer splitter plate; V-shaped fairings spilling the boundary layer connect the splitter plates to the fuselage.

The centre fuselage is area-ruled, narrowing markedly at the wing/fuselage joint and widening again aft of the wings; it incorporates S-shaped inlet ducts serving the engines, with fuel tanks in between. The rear fuselage is detachable for engine maintenance and change, accommodating the engines. It incorporates four airbrakes located in a cruciform arrangement and a brake parachute bay located ventrally between the engine jetpipes.

Wings and tail unit: The wings and tail unit are identical to those of the Su-9/Su-11, except for the changes to the wings at the root to match the area-ruled fuselage and the addition of two boundary layer fences.

Landing gear: Hydraulically retractable tricycle type, with single wheel on each unit; the nose unit retracts forward, the main units inward into the wing roots. All three landing gear struts have levered suspension and oleo-pneumatic shock absorbers. The nose unit is equipped with a 600 x 155 mm (23.6 x 6.1 in) KT-104 brake wheel and a shimmy damper; the main units have 880 x 230 mm (34.6 x 9.0 in) KT-69/4 wheels.

Powerplant: Two Tumanskiy R-11F2S-300 axial-flow afterburning turbojets rated at 3,900 kgp (8,600 lbst) at full military power and 6,175 kgp (13,610 lb st) in full afterburner. (For a detailed description of this engine, see Chapter 7.) Provisions were made for installing R-11F2SU-300 engines features bleed valves for the flaps' boundary layer control system.

Fuel system: Internal fuel is carried in four integral tanks – two fuselage tanks and two wing tanks. The total capacity is 5,120 litres (1,126 Imp gal) or 4,200 kg (9,260 lb). There are two 'wet' hardpoints under the centre fuselage permitting carriage of two 600-litre (132 Imp gal) drop tanks increasing the total to 6,320 litres (1,390 Imp gal) or 5,180 kg (11,420 lb).

Armament: Two K-8M2 medium-range air-to-air missiles with semi-active radar homing or IR homing carried on pylon-mounted PU-1-8 launch rails, one under each wing, as on the Su-11.

Avionics and equipment: The avionics suite includes an RSIU-5 (R-802V) VHF communications radio, an ARK-10 ADF, an RV-UM low-range radio altimeter, an MRP-56P marker beacon receiver, SOD-57M DME, a Lazoor' (ARL-S) command link receiver working with the Vozdookh-1M GCI system, an SRZO-2M Kremniy-2M IFF transponder and (provisionally) a Sobol' fire control radar.

Su-15 production interceptor
(T-58, *izdeliye* 37 or *izdeliye* 37 Srs D)

On 30th April 1965 the Council of Ministers issued a directive formally including the Su-15-98 aerial intercept weapons system into the PVO inventory. Thus the aircraft officially received the service designation Su-15 (ie, it was recognised as a new type, not a 'modernised Su-11'), while the K-98 missile was redesignated R-98. The directive required the Novosibirsk aircraft factory No.153 to launch series production of the new interceptor in early 1966.

The Sukhoi OKB had completed detail design work on the production version of the interceptor incorporating the changes requested by the State commission

Three excellent photos of the first pre-production Su-15 ('34 Red', c/n 0015301). Details worthy of note include the outward-canted intake trunks, the fuselage 'waist' in that area and the anti-glare panels.

Above: Su-15 c/n 0015301 at the Sukhoi OKB's premises at Moscow-Khodynka at a later date, wearing the new tactical code '01 Red'. Note the black/yellow 'Danger, air intake' markings and the open access panel.

(increased fuel capacity etc.) back in the second half of 1964, transferring the blueprints to the Novosibirsk aircraft factory. In the course of 1965 the plant, which had been building the Yak-28P interceptor since the termination of Su-11 production, geared up to build the Su-15 which received the product code *izdeliye* 37. (Later, when the Su-15T/Su-15TM entered production, the original model was sometimes referred to in official documents as *izdeliye tridtsat' sem' serii D* (product 37 Srs D) in order to differentiate it from the new version, or *izdeliye tridtsat' sem' serii M*; the D probably referred to the original designation T-58D.)

Construction of the first two aircraft in the pre-production batch (Batch 00) began at the end of the year. As already mentioned, initially the Su-15 had no chief project engineer; it was not until it had entered production that OKB-51 engineer N. P. Polenov was assigned overall responsibility for this aircraft.

Aircraft designers are only human, and one can hardly blame them for wishing to promote their own products – unless they resort to dirty tricks, of course. Alarmed by the steady progress of the Su-15's State acceptance trials and perceiving it as a threat to the Yak-28P (to which the new Sukhoi interceptor was clearly superior), OKB-115 General Designer Aleksandr S. Yakovlev urgently set about developing a modernised version of his interceptor with similar air intakes. Actually 'similar' is a mild way of putting it; in fact, the chief of OKB-115 urgently despatched his elder son Sergey A. Yakovlev to Novosibirsk to have a close look at the section of the Su-15's manufacturing drawings referring to the air intakes. Yet the radically but hastily redesigned Yak-28-64 (alias Yak-28N) featuring buried engines and Su-15 style two-dimensional lateral intakes turned out to be a lemon – the first few test flights revealed it was inferior to the standard Yak-28 with underwing engines, never mind the Su-15! The programme was abandoned and the Sukhoi OKB regained control over the Novosibirsk plant. 'The empire strikes back'?

Building the Su-15 did not involve a complete change of manufacturing technology, since the aircraft had considerable structural commonality with the Su-9 and Su-11 which the factory had built earlier; it also had a lot in common with both the Su-11 and the Yak-28P as regards systems and equipment. Yet the first pre-production example (c/n 0015301; the abbreviated form 0001, as stencilled on the actual aircraft, is sometimes used) took a long time to complete; coded '34 Red', it was rolled out at Novosibirsk-Yel'tsovka on 21st February 1966, several months late. On 6th March the aircraft made its first flight with factory test pilot I. F. Sorokin at the controls; in late April it was ferried to the Sukhoi OKB's flight test facility in Zhukovskiy. The second pre-production Su-15 (c/n 0015302) was completed in June 1966, joining the first aircraft on 21st July.

As mentioned earlier, in early 1965 the OKB had tested new cranked delta wings on the first prototype. However, the factory refused to build the fighter with this wing design on the pretext that no verdict from the 'ultimate authority' (GK NII VVS) existed on the soundness of this design. In reality the reason was more down to earth but nonetheless plausible: the jigs and tooling for the Su-15's original delta wings had already been manufactured; modifying them would mean further delays in the production schedule, which was in jeopardy as it was.

Full-scale production of the Su-15 gradually gained momentum from mid-1966 onwards; 17 production aircraft had been assembled by the end of that year. The production model featured an internal fuel capacity of 6,860 litres (1,509 Imp gal) or 5,600 kg (12,345 lb) thanks to the constant-width centre fuselage patterned on the post-modification T58D-3 but the number of fuselage tanks was reduced to three. Production aircraft were powered by R11F2S-300 (*izdeliye* 37F2S) engines rated at 3,900 kgp (8,600 lbst) dry and 6,175 kgp (13,610 lb st) reheat. The SAU-58 automatic flight control system, on the other hand, took some time coming, passing its State acceptance trials only in 1968. The trials revealed that the system required major changes and it was decided to postpone its introduction until the aircraft's next upgrade. Still, the installation of the SAU-58 had been planned from the outset and space for its modules was reserved in the avionics bays. As a result, the aircraft had neither the AFCS nor the simpler AP-28 autopilot fitted to the prototypes; the road to hell is paved with good intentions! Worse, the production Su-15 did not even have dampers in any of the control circuits, the ARZ-1 artificial-feel unit in the tailplane control circuit being the only stability augmentation system. Curiously, the military never demanded changes in this area, as the stability and handling characteristics of production aircraft during pre-delivery tests appeared adequate.

The second production Su-15 ('67 Red', c/n 0115302) seen during trials. Note the –o– photo calibration markings on the forward and rear fuselage. This aircraft was used to investigate a problem associated with abruptly increasing stick forces.

'69 Blue', an early-production Su-15 (c/n 031506).

Right from the start the production aircraft featured an improved version of the air intake control system designated UVD-58M and the KS-4 ejection seat permitting safe ejection throughout the flight envelope, including take-off and landing, providing the speed was above 140 km/h (87 mph).

The avionics fit of early-production Su-15s included an RSIU-5 (R-802V) two-way VHF communications radio, an MRP-56P marker beacon receiver, an RV-UM low-range radio altimeter, an ARK-10 ADF, SOD-57M distance measuring equipment, a Lazoor' (ARL-S) command link receiver, an SRZO-2M IFF transponder, a Sirena-2 radar warning receiver, a KSI-5 compass system and an AGD-1 artificial horizon. Production aircraft were equipped with the Oryol-D58 radar which, in its production form, was officially designated RP-15. Increasing the radar's resistance to jamming and perfecting the missiles' proximity fuses was the responsibility of other bureaux. This, too, proved to be a protracted affair and the trials of the improved RP-15M (Oryol-D58M) radar held jointly by the manufacturer and the Air Force were not completed until 1967 when the Su-15 had become operational. New-build Su-15s were henceforth completed with the modified radar, while previously built examples were upgraded *in situ* to the new standard.

A typical production Su-15 had an empty weight of 10,220 kg (22,530 lb). The production rates built up gradually; in 1968 the factory turned out 150 Su-15s versus 90 in 1967.

Between 7th July 1967 and 25th September 1968 GK NII VVS held checkout trials of a Batch 2 Su-15 (c/n 0215301) to see if the performance figures matched those obtained during the State acceptance trials. The results were disappointing, and the causes were traced to two main problems. The first problem lay with the engines. The R11F2S-300 engine was produced by two factories, No.500 in Moscow and No.26 in Ufa, Bashkiria. In the course of the tests Su-15 c/n 0215301 flew consecutively with three sets of engines, two of which were manufactured in Moscow and the third in Ufa; the Moscow-built engines provided the required performance but the Ufa-built ones did not (this concerned first and foremost interception range in pursuit mode).

Between 10th June and 30th August 1969 the same aircraft underwent renewed tests with a new pair of Ufa-built engines. This time the performance improved to an acceptable level, except that the interception range in pursuit mode against a target doing 1,400 km/h (870 mph) at 23,000 m (75,460 ft) turned out to be 25 km (15.5 miles) less than required. GK NII VVS immediately solved the problem by a bit of trickery: the next mission was flown with a single drop tank instead of two to reduce drag and the required interception range of 195 km (121 miles) was obtained.

The other nasty surprise was that in certain flight modes the stick forces would suddenly increase abruptly – the stick seemed to bump into invisible stops. The problem was traced to the hydraulic control surface actuators which were not powerful enough. The OKB and the military turned to LII for help; at the end of 1968 the institute held a special research programme which was performed by test pilot A. A. Shcherbakov in the fourth production Su-15 (c/n 0115304). Concurrently the OKB investigated the problem on its own, using the second pre-production Su-15 (c/n 0015302) and the second production aircraft ('67 Red', c/n 0115302), but soon pressure of higher-priority programmes caused the issue to be put on hold.

The Su-15 and its systems underwent constant refinement in the course of production. The biggest number of upgrades was made in 1968 when the new wings with the cranked leading edge and the BLC system were introduced after all. The first shipset of R11F2SU-300 (*izdeliye* 37F2SU) engines was delivered to OKB-51 in early 1968; the U in the designation stood for [*sistema*] *oopravleniya* [*pogranichnym sloyem*] – boundary layer control system. After that, the first pre-production Su-15 was re-engined and rewinged. After a brief initial flight test programme performed by OKB pilots the fighter was flown to the GK NII VVS facility in Akhtoobinsk to undergo more extensive testing. The military test pilots observed that the extended-chord outer wing portions had a positive effect on the fighter's stability. On the minus side, the double-delta wings caused a slight reduction in performance at supersonic speeds and a reduction of the service ceiling; yet there was no cause for alarm, since the performance figures obtained were still within the limits specified by the military.

Regarding the blown flaps, the Air Force test pilots noted that the BLC caused a forward shift in the aircraft's CG, which complicated the landing procedure because the stabilator travel limits were insufficient; on the other hand, the 40 km/h (25 mph) reduction in approach/landing speed was an undoubted asset. The final conclusion was a thumbs-up and the modifications were recommended for incorporation on the production fighter.

In 1969 the blown flaps were introduced on production Su-15s from c/n 1115301 onwards but the old pure delta wings were retained for the time being; the first production Su-15 to be completed with double-delta wings was c/n 1115331. From c/n 1115336 onwards all Su-15s had provisions for installing the new R13-300 engines. Another change introduced on batch 11 was the addition of a 'black box' – an SARPP-12V-1 flight data recorder (FDR).

As was customary, all production Su-15s underwent pre-delivery tests; these showed that the modified wings reduced the interceptor's service ceiling by an average 400 m (1,310 ft). As a result, the PVO temporarily stopped accepting new Su-15s; Pavel O. Sukhoi's aide Yevgeniy A. Ivanov, who was responsible for the Su-15 programme, received a dressing-down from the 'head office' at GKAT. Several Batch 11 aircraft were urgently delivered to GK NII VVS for the purpose of holding checkout trials, and a Sukhoi OKB delegation flew to Akhtoobinsk to attend these trials. The trials confirmed the reduction; yet both the military and the OKB were well aware that nothing could be done to increase the service ceiling. Hence a protocol of agreement was signed in which the PVO conceded that the service ceiling of the rewinged Su-15 would be 18,100 m (59,380 ft) versus 18,500 m (60,695 ft) for the original 'pure delta' version.

The issue of the 'invisible stops' hampering stick travel was addressed again after the cranked-wing version had been put into production, and the remedy was to install more powerful hydraulic actuators. Tests held in 1970 showed that on aircraft featuring the new BU-220 actuators the phenomenon recurred only when the airbrakes were deployed or if one of the control system's hydraulic circuits failed. From then on identical BU-220 actuators were installed in all three control circuits on production Su-15s.

The powerplant and some other systems also had their share of bugs. While being more reliable than the AL-7F, the R11F2S-300 engines powering the Su-15 were not immune against surging or flaming out at high altitude, and relighting them after a flameout was not always easy. Actual Su-15 operations in the PVO's first-line units added new details to the picture. Among other things, it turned out that the engines were prone to catching fire, while the fire warning and fire suppression systems were unreliable. In 1968-69 alone, two accidents and six incidents (in the Russian terminological sense – ie, near-accidents or events not grave enough to be rated as a non-fatal accident) occurred in PVO fighter units for the reasons stated above.

Sukhoi OKB test pilots also found themselves in such dangerous situations. Thus on 30th September 1969, when Vladimir S. Ilyushin was making yet another flight in Su-15 c/n 0015302 with the objective of testing the SAU-58 AFCS, he discovered that the nozzle petals on one of the engines would not close when the afterburners were switched off. He chose to abort the mission and return to base urgently – and not a minute too soon, as it turned out. Examination of the failed engine showed that the afterburner chamber had burned out near one of the flame holder's attachment points. The escaping jet of hot gases destroyed the hydraulic lines feeding the nozzle actuators and started burning through the rear fuselage structure; at the moment of touchdown the rear fuselage mainframe was a hair's breadth away from failing catastrophically. On 12th June 1972 a similar failure resulted in the crash of Su-15 c/n 1115342 – incidentally, the sole 'company-owned' example to be lost in an accident. Test pilot A. S. Komarov was making a routine flight to establish the aircraft's roll stability characteristics and measure the temperature inside the rear fuselage. He was less lucky than Ilyushin; when the afterburner failed the fire burned through the control runs, rendering the machine uncontrollable and leaving no choice but to punch out. Komarov sustained serious injuries in the ejection that kept him grounded for two years before the doctors decided that he was fit to fly again.

Production of the initial version with no suffix letters to the designation (or 'Su-15 *sans suffixe*') peaked in 1969 when 165 examples were built. In 1971 the Su-15 *sans suffixe* was succeeded on the Novosibirsk production line by the more advanced Su-15TM.

The production-standard Su-15 made its public debut on 9th July 1967 at the grand air show at Moscow-Domodedovo airport. After that the ASCC allocated the reporting name *Flagon* to the new interceptor; this was later changed to *Flagon-A* when new versions became known.

A number of production Su-15s was used in various research and development programmes in 1966-75 – mostly as avionics and equipment testbeds, verifying items which found use on later versions of the Su-15. Twenty-six such R&D programmes were completed in 1966-96 alone. For instance, in 1969 and 1971 the first pre-production

Two photos of the T-58L (or T58-L) development aircraft in late configuration; it was converted from the T58D-2, retaining the latter's tactical code '32 Blue'. The new twin-wheel nose gear unit giving the aircraft a marked nose-up attitude is clearly visible. The inscription on the tail reads 58-Л (58-L in Cyrillic characters).

example (c/n 0015301) was used to test new shapes of the fairing between the engine nozzles; in 1970 the same aircraft served to determine the field performance with the engines in minimum afterburner mode. Some other development aircraft are listed below.

T-58L (T58-L) development aircraft

In the early 1960s, when the Cuban missile crisis put the two superpowers on the brink of an all-out war, the Soviet Armed Forces paid much attention to dispersing troops (including Air Force units) in order to make them less vulnerable to enemy strikes. In the case of military aviation this meant operations from unpaved tactical and reserve airstrips – for which, as it turned out, the units of the VVS and the IA PVO were totally unprepared. This led the military to demand insistently that all tactical aircraft types should be capable of operating from semi-prepared dirt strips.

After the State acceptance trials the Sukhoi OKB followed the recommendations of the State commission, developing a special skid landing gear for the second prototype Su-15 (T58D-2) with a view to exploring the possibility of operating the new interceptor from unpaved strips. By then the OKB had accumulated a wealth of experience with the S22-4, S-23 and S-26 development aircraft, all of them versions of the Su-7B featuring different versions of skid landing gear. Thus the optimum layout could be chosen and the design work completed within a short time frame. The engineers selected a mixed arrangement with skids on the main units and a wheeled nose unit, as on the S-26.

Modification work on the T58D-2 was completed in the first half of 1965, whereupon the aircraft was redesignated T-58L (the designation is sometimes rendered as T58-L), the L standing for *lyzhnoye shassee* – skid landing gear. The conversion involved installing new main gear units which could be quickly reconfigured from wheels to skids and back again, with appropriate modifications to the mainwheel wells and main gear doors; the skids were provided with a lubrication system to facilitate movement on grass and packed earth surfaces. The standard castoring nose gear unit was replaced by a steerable unit, as on the S-26 (normally the T-58 was steered on the ground by differential braking, which was impossible with skids). Provisions were made for installing jet-assisted take-off (JATO) solid-fuel rocket boosters, changes were made to the forward fuselage and cockpit, and a new KS-4 ejection seat was fitted.

Vladimir S. Ilyushin performed the T-58L's first flight on 6th September 1965. From 1966 until the mid-1970s the aircraft underwent extensive testing on various semi-prepared grass, dirt and snow strips in various climatic zones; GK NII VVS also participated in these tests. The aircraft was also flown by OKB test pilots Ye. K. Kukushev, Ye. S. Solov'yov, V. A. Krechetov *et al*., as well as by GK NII VVS pilots.

In one of the test flights with Kukushev at the controls the aircraft banked at the moment of rotation and the anti-flutter boom on one of the stabilators dug into the ground, ripping away together with a portion of the skin. To prevent a repetition of this incident the anti-flutter booms were angled up 15° on all Su-15s, production aircraft and prototypes alike. Shortly afterwards the T-58L was refitted with a new, taller nose gear unit featuring twin 620 x 180 mm (24.4 x 7.0 in) KN-9 non-braking wheels; the purpose of this modification was to increase the angle of attack on take-off (thereby increasing lift and shortening the take-off run), raise the air intakes higher above the ground (thereby reducing the risk of foreign object ingestion) and improve ground manoeuvrability.

The skid landing gear was not introduced on the production model, as the tests of the T-58L revealed major operational problems associated with operations from unpaved strips. The vibrations experienced on uneven runways subjected the avionics and armament to augmented loads which could ruin them; also, the missiles were liberally spattered with dirt, which likewise could put them out of action. It should be noted that production Su-15s with a standard landing gear also underwent tests on unpaved runways; these tests ultimately led to the introduction of a twin-wheel steerable nose gear unit on the Su-15TM.

Upon completion of the test programme in 1974 the T-58L was donated to the Air Force Academy named after Nikolay Ye. Zhukovskiy as an instructional airframe. Fortunately it was not 'vivisected', as ground instructional airframes often are, and subsequently moved to the Soviet Air Force Museum in Monino.

Su-15/R11F3-300 engine testbed

The fourth production Su-15 (c/n 0115304) was used by LII as a testbed for the Tumanskiy R11F3-300 engine featuring a contingency rating. Actually this term was used in a non-standard way, as it referred to increasing thrust at low altitudes to improve performance, not to an automatic power reserve activated in the event of an engine failure.

T-58-95 engine testbed

In keeping with MAP and Air Force orders issued in October 1967 and May 1968 respectively several production Su-15s were modified for testing the new Gavrilov R13-300 afterburning turbojets. The R13-300 (*izdeliye* 95) was a derivative of the R13F-300 developed by the Ufa-based *Soyuz* (Union) engine design bureau led by S. A. Gavrilov. At that time the young OKB had little design experience of its own and thus ran into major problems with the R13-300 which differed significantly from the precursor; among other things, the number of high-pressure compressor stages was increased from three to five and a second afterburner stage was added. Thanks to these changes the engine delivered 4,100 kgp (9,040 lbst) at full military power and 6,600 kgp (14,550 lb st) in full afterburner versus 3,900 kgp (8,600 lbst) and 6,175 kgp (13,610 lb st) respectively for the R11F2S-300.

Upon completion of the bench tests a flight-cleared R13-300 was installed on Su-15 c/n 0415302 by August 1967, replacing the starboard R11F2S-300; the resulting 'lop-sided' aircraft was known at the Sukhoi OKB as the T-58-95, the last two digits referring to the development engine's product code. Suitably fitted out with data recording equipment, the T-58-95 made 11 flights under the initial flight test programme before being transferred to LII where the manufacturer's tests of the R13-300 engine were held in 1967-68.

Su-15/R13-300 engine testbeds

By mid-December 1968 the Sukhoi OKB had installed a complete shipset of R13-300 engines in a production Su-15 coded '11' (c/n 0715311). After completing manufacturer's flight tests in March 1969 the aircraft was flown to the GK NII VVS facility in Akhtoobinsk for State acceptance trials of the new engine and the re-engined interceptor as a whole. The trials programme involving 53 flights showed that the service ceiling, acceleration time, effective range, combat radius and field performance had improved, thanks to the new engines.

Between November 1969 and February 1970 Su-15 c/n 0715311 underwent additional testing to explore the R13-300's resistance to surging; the test programme was performed by Vladimir S. Ilyushin. Another Su-15 coded '37' (c/n 1115337), an aircraft owned by LII and one of the first examples to be powered by R13-300 engines, joined the test programme in December 1969. In the course of the tests engine flameouts at high Mach numbers were experienced on both aircraft when the engines were running in full afterburner. It turned out that the engines were simply not getting enough air because the air intakes were too small. To eliminate this dangerous phenomenon the air intakes had to be widened but this was impracticable in mass production at the time, since it required major structural changes and hence changes to the tooling. Hence it was decided simply not to engage the second afterburner stage on Su-15s *sans suffixe* equipped with R13-300 engines, since the first afterburner

Above: Su-15 '16 Blue' (c/n 0615316) was used by the Sukhoi OKB as a weapons testbed. This photo shows the test equipment pods on the wing pylons (including an AKS-5 camera pod) and the photo calibration markings on the fuselage and tail.

Left: This head-on view shows the non-standard outer wing pylons.

Below left: Here, '16 Blue' carries an unidentified munition under the fuselage.

nosti), aka LNPO Leninets (Leninist), converted the abovementioned Su-15 coded '11' (c/n 0715311) into an avionics testbed. (LNPO = *Leningrahdskoye naoochno-proizvodstvennoye obyedineniye* – Leningrad Scientific & Production Association. This company, which was one of the Soviet Union's leading avionics houses, is now the Leninets Holding Company.) The Oryol-D58M fire control radar was replaced by a prototype *Rel'yef* (Profile) terrain following radar developed for the T-6 (Su-24) tactical bomber; this radar enabled automatic terrain-following flight during low-level air defence penetration. The aircraft was designated SL-15R or T-58R, the SL standing for *samolyot-laboratoriya* (laboratory aircraft) in keeping with LNPO Leninets's system of designating its avionics testbeds and the R suffix referring to the Rel'yef radar.

Outwardly the SL-58R was no different from any production Su-15, since the experimental radar was housed within the standard radome. The aircraft was equipped with a KT-61/3 nosewheel and KT-117 mainwheels.

stage provided an adequate increase of flight performance.

Nevertheless a third production Su-15 (identity unknown) was re-engined with R13-300s, undergoing tests in 1970-71. This aircraft featured wider air intakes to cater for the engines' greater mass flow; this intake design, together with the R13-300 engines, was later introduced on the ultimate Su-15TM.

Su-15 ECM/IRCM equipment testbed

From 1969 onwards LII used a specially equipped Su-15 (identity unknown) to study the reflection of radar echoes from targets, clouds of chaff and the ground. Later this aircraft was further modified as a testbed for passive electronic countermeasures (ECM) and infra-red countermeasures (IRCM) equipment, serving to verify almost all chaff/flare dispenser types used by the Soviet/Russian Air Force.

SL-15R (T-58R) avionics testbed

In May 1972 the Leningrad-based NII-131 of the Ministry of Electronics Industry (MRP – *Ministerstvo rahdioelektronnoy promyshlen-*

Su-15 communications equipment testbed

Another Su-15 (identity unknown) was used in 1968-69 to test the new R-832M Evkalipt-SM (Eucalyptus-SM) communications radio.

Su-15 aerodynamics research aircraft for spinning tests and uncontrollable roll research

Two specially modified Su-15s were used for extensive aerodynamics research to investigate the aircraft's stability and control characteristics in certain flight modes. Thus the

Flight Research Institute twice held spinning tests of the Su-15 – in 1968, using an unidentified early-production aircraft with pure delta wings, and in 1973, using the second production example of the double-delta version ('37', c/n 1115337). Oleg V. Goodkov was project test pilot in both cases, but Igor' P. Volk performed part of the test programme on c/n 1115337. A while earlier, in 1970, this aircraft was used to check the flight performance in minimum afterburner mode; in 1971 it was used to verify landing techniques developed for a fully forward-CG configuration. It was not until 1972 that Su-15 c/n 1115337 was equipped with PPR-90 spin recovery rockets (*protivoshtopornaya porokhovaya raketa*), enabling high-alpha and spinning tests to be held. A similar spinning test programme was performed by GK NII VVS, with N. V. Kazarian as project test pilot.

The tests showed that considerable vibration set in when the aircraft approached critical AOAs, warning the pilot that he was 'pushing it too far'. Actually the Su-15 could enter a spin only due to a grave piloting error or if the spin was initiated intentionally. The spinning characteristics of the two versions were similar, but the double-delta version was more stable during the spin.

Another test programme was held in response to a series of accidents in which the aircraft suddenly started rolling uncontrollably during vigorous manoeuvres at supersonic speeds, the pilots perceiving this as a critical control system failure. The second production Su-15 (c/n 0115302), which still had a test equipment suite from an earlier trials programme, was set aside to investigate into the problem. In 1970-71 the Sukhoi OKB and LII held a joint test programme to determine the conditions in which this phenomenon occurred – first at subsonic speeds and then beyond Mach 1; Yevgeniy S. Solov'yov and Igor' P. Volk were the OKB's and the institute's project test pilots respectively. It was established that there were no control system failures at all – the cause was traced to a specific relationship of inertia forces along different axes; the pilots provoked the uncommanded spin by pulling negative g. The correct course of action in this situation was to reduce speed and set the controls to neutral.

T-58K (Su-15K) aerodynamics research aircraft with modified wings

In 1968 the Sukhoi OKB started work on modified wings featuring a sharp leading edge. By April 1973 this work reached the practice stage – the fourth production Su-15 (c/n 0115304) was converted into the T-58K research vehicle (the K stood for [*modifitseerovannoye*] *krylo* – [modified] wings). The boundary layer fences and missile pylons were deleted, the pylon attachment points being enclosed by special fairings, and a new extended leading edge section with a more pointed profile was installed. The BU-220 hydraulic actuators in the tailplane control circuit were replaced by BU-250 units, and part of the standard avionics was replaced by test equipment and ballast. The T-58K underwent trials at LII in 1973-74.

Su-15 cannon armament testbeds

Originally the Su-15's armament consisted solely of two air-to-air missiles; yet the military kept requesting that cannon armament be incorporated as well. At first the Sukhoi OKB intended to use a single GP-9 standardised centreline pod housing a 23-mm (.90 calibre) Gryazev/Shipoonov GSh-23 twin-barrel cannon (GP = *gondola pushechnaya*, which translates (and is very conveniently deciphered) as 'gun pod'). After a series of tests this was found suitable for the Su-15, and the Novosibirsk aircraft factory even built ten Batch 12 aircraft with the appropriate attachment fittings and connectors. However, no GP-9 pods were in fact ever delivered to Su-15 units – a strange circumstance, considering that the VVS's tactical fighter units equipped with the MiG-21S (which, too, was originally armed only with two AAMs) received these pods.

By then, however, the military had changed their requirements; the UPK-23-250 pod containing the same cannon but with 250 rounds instead of 200 became the Su-15's standard cannon armament. In addition to the greater ammunition supply, the UPK-23-250 could be fitted and removed extremely easily (unlike the GP-9); most importantly, two such pods could be carried on the Su-15's fuselage hardpoints instead of drop tanks, giving twice the firepower.

A production Su-15 (c/n 1115342) was set aside for cannon armament tests, passing its State acceptance trials in March-September 1971. Even though aiming accuracy with the Su-15's standard K-10T sight left a lot to be desired, the cannon pod installation was recommended for use against both air and ground targets. Since production of the basic Su-15 *sans suffixe* had ended by then, these aircraft were retrofitted with UPK-23-250 pods in service.

Su-15 with wing cannons (project)

In 1968-69 the Sukhoi OKB considered the idea of equipping the Su-15 with a built-in cannon. A cannon bay was envisaged in the starboard wing root; it was to house an experimental 23-mm cannon developed by the Fine Machinery Design Bureau (*Konstrooktorskoye byuro tochnovo mashinostroyeniya*) in Tula, a city renowned for its armourers. Provisionally designated *izdeliye* 225P, this was a fairly compact Gatling cannon with a high rate of fire. Yet accommodating the cannon's ammunition supply inside the Su-15's thin wing proved to be an insurmountable task and the idea was dropped.

T-58VD experimental STOL aircraft

In early 1965, when the T58D-1 had completed a brief flight test programme with the new double-delta wings, the OKB decided to use this aircraft as a propulsion systems testbed and a STOL technology demonstrator in conjunction with the development of the T-58M low-altitude attack aircraft. (The latter designation proved to be shortlived; the T-58M, which later became the T-6, was a totally unrelated design and a much larger aircraft which evolved into the Su-24 tactical bomber and lies outside the scope of this book.) This involved installing small turbojet engines vertically inside the fuselage to generate lift. The lift-jet concept was quite popular then both in and outside the Soviet Union. Western development aircraft making use of lift engines included the Dassault Balzac supersonic fighter and the Short SC.1 technology demonstrator, although both of them were vertical take-off and landing (VTOL), not STOL, aircraft.

On 6th May 1965 MAP issued an order requiring the Sukhoi OKB to build and test a proof-of-concept vehicle in order to verify the STOL technology using lift-jets. The engineers set to work; by mid-year they had completed the project documents for the conversion of the T58D-1 into such a vehicle. The extensive conversion involved remanufacturing the centre fuselage to accommodate three 2,350-kgp (5,180-lbst) RD36-35 turbojets developed by the Rybinsk-based OKB-36 under Pyotr A. Kolesov. The lift engines were installed in a bay between the cruise engines' inlet ducts with the axes inclined forward 10°, breathing through two large scoop-type intakes on the fuselage's upper surface (the forward intake served the foremost engine and the rear intake the other two). The exhaust aperture was closed by louvres which had two operational settings, directing the jet exhaust aft on take-off to add a measure of forward thrust or forward on landing to slow the aircraft down. In cruise flight (in 'clean' configuration) the lift-jets' intakes and exhaust louvres closed flush with the fuselage skin. The conversion also included modifications to the wings, relocation of various equipment and piping and the like.

All fuel was now carried in the wing tanks – and that means less fuel and three more engines guzzling away at it. But then, range and endurance were not crucial for a pure technology demonstrator that was not meant to operate far away from its base.

Designated T-58VD (for *vertikahl'nyye dvigateli* – 'vertical engines', ie, lift-jets), the

Above: This view of the T-58VD experimental STOL aircraft shows the dorsal air intake scoops for the three lift engines and the modified double-delta wings, as well as the photo calibration markings.

Left: The T-58VD on a tethered test rig for checking the effect of the lift engines, with a high-power blower ahead of the aircraft. The stencil on the tail reads '58ВД' ('58VD' in Cyrillic). The shed conceals the aircraft from US surveillance satellites.

rebuilt aircraft was completed at the end of 1965, commencing tethered tests on a purpose-built ground rig on the Sukhoi OKB's premises; R. G. Yarmarkov retained his assignment as this aircraft's engineer in charge of the tests. The rig featured an 'open-air wind tunnel' – a Kuznetsov NK-12 turboprop engine driving ducted propellers emulated the slipstream at simulated speeds up to 400 km/h (248 mph), creating proper operating conditions for the lift engines. The pad on which the aircraft sat was rigged with pressure sensors to capture the off-loading of the landing gear and thus assess the efficiency of the lift engines.

The tethered tests allowed the T-58VD's aerodynamics with the lift engines running to be explored and the operation of all principal systems to be checked. Unfortunately they were marred by a tragic incident in February 1966. A mechanic from the OKB's propulsion laboratory was careless enough to approach the aircraft's forward fuselage when the blower was running; the powerful stream of air immediately swept him off his feet and hurled him savagely against the aircraft, killing him.

On 26th April 1966 upon completion of the ground test phase the aircraft was trucked to LII, making its first flight on 6th June at the hands of Yevgeniy S. Solov'yov. Later on, the T-58VD was flown by both Solov'yov and Sukhoi OKB chief test pilot Vladimir S. Ilyushin; by the end of the year it had made 37 real flights and 19 taxi runs and short hops, including high-speed runs on a dirt strip. The manufacturer's flight tests showed that the lift created by the auxiliary engines reduced the unstick speed from 390 to 285 km/h (from 242 to 177 mph) and the landing speed from 315 to 225 km/h (from 195 to 139 mph). The take-off run was shortened from 1,170 to 500 m (from 3,840 to 1,640 ft) and the landing run from 1,000 to 560 m (from 3,280 to 1,840 ft) – an impressive result. On the other hand, it became apparent that the chosen location of the lift engines was not the optimum one, as the thrust of the forward engine caused a strong tendency to pitch up during landing approach; the problem was solved by using only the centre and rear lift-jets for landing.

On 9th July 1967 the T-58VD participated in the air show at Moscow-Domodedovo, giving a short take-off and landing demonstration with Yevgeniy S. Solov'yov at the controls. After that, the STOL version received the reporting name *Flagon-B*.

The results of the T-58VD's flight tests gave the Sukhoi OKB valuable experience in designing, building and testing STOL aircraft and allowed the test pilots to master the technique of flying such aircraft. This knowledge was incorporated into the design of the delta-wing T6-1 strike aircraft prototype. However, the trade-off for the good short-field performance turned out to be too high; the lift engines reduced the space available for fuel dramatically while significantly increasing fuel consumption on take-off and landing. Also, the operation of the lift-jets worsened longitudinal stability somewhat, and in cruise mode the lift engines were just a lot of useless weight which reduced the payload. Hence the second prototype, the T6-2I, was radically reworked to feature variable-geometry wings which gave the desired results.

Upon completion of the test programme the T-58VD was donated to the Moscow Aviation Technology Institute (MATI) where it served as an instructional airframe for a while.

Su-15 CCV with a side-stick

In 1980 the Sukhoi OKB converted a production Su-15 (c/n 1115328) into a control con-

Right, below and bottom: The T-58VD as it appeared at the 1967 Domodedovo air show. The tactical code and the '58ВД' tail titles are gone and the nose is painted differently. Note the twin aerials under the nose. The bulged fairings replacing the upper airbrake panels probably house test equipment

figured vehicle (CCV) with variable in-flight stability and controllability parameters. For the first time in Soviet aircraft design practice the aircraft featured a side-stick controller; the standard centrally mounted stick was retained and the pilot was able to switch the control system from one stick to the other as required.

The CCV underwent tests at LII between 1981 and 1982; it was flown by LII test pilots V. I. Loychikov, Rimas A. A. Stankiavicius, A. S. Lev-chenko, Igor' P. Volk, A. S. Shchookin, V. V. Zabolotskiy and Yuriy A. Oosikov. Unfortunately on 11th November 1982 the aircraft crashed and was destroyed before the test programme could be completed. Oosikov ejected and survived but sustained serious injuries which forced him to give up flying.

Su-15 multi-role testbed (c/n 1315340) ('aircraft 0009')

As already mentioned, tests on R13-300 engines on Su-15 c/n 0715311 revealed a tendency to flame out at high Mach numbers because the air intakes were too small. Hence in 1970 the Sukhoi OKB modified a further production Su-15 (c/n 1315340) to feature widened air intakes which were later incorporated on the production Su-15TM. For security reasons this aircraft was referred to at the OKB as 'aircraft 0009' (possibly with a view to fooling would-be spies into thinking it was the aircraft's c/n).

After fairly lengthy manufacturer's tests the aircraft was turned over to GK NII VVS, making 44 test flights there between mid-May and August 1971 as part of the Su-15TM's State acceptance trials. The aircraft showed encouraging performance in comparison with the Su-15TM prototype (c/n 0115305) which featured narrow intakes (see later); acceleration at supersonic speeds was improved, the service ceiling increased from 17,600 to 18,500 m (from 57,740 to 60,695 ft) and effective range increased to 1,680 km (1,040 miles).

In June 1972 'aircraft 0009' was used to test the new ogival radome developed for the Su-15TM.

Su-15 missile armament testbeds

When the Su-15TM armed with R-98M missiles entered service it was decided to adapt existing Su-15s *sans suffixe* equipped with the Oryol-D58 radar for using this missile which was originally designed for use with the Taifoon radar. To this end a late-production Su-15 (c/n 1415301) was converted in 1975, successfully passing a special test programme.

By the early 1970s the **Mol**niya (Lightning) OKB had completed development of the R-60 agile short-range AAM and the Air Force selected it as the main close-in weapon for all Soviet fighter types. After the R-60 had been tested on the Su-15TM it was decided to arm the Su-15 *sans suffixe* with this missile as well. A suitably modified early-production Su-15 (c/n 0615327) was tested successfully in 1978-79, the Air Force's aircraft overhaul plants started upgrading Su-15s with launch rails for two R-60s. A configuration with four R-60s on APU-60-2 paired launchers (*aviatsionnaya pooskovaya oostanovka* – aircraft-mounted launcher) was also tested but did not find its way into service.

Top left: The T-58VD climbs away with the landing gear already retracted and the lift engines running.

Centre left: The T-58VD makes a short take-off from an unpaved airstrip, the lift jets kicking up a minor dust storm.

Lower left: This close-up of the T-58VD's centre fuselage with all engines running shows the open dorsal intake scoops and the auxiliary blow-in doors of the cruise engines.

Bottom left: The T-58VD in its latter days as an instructional airframe at the Moscow Aviation Technology Institute, with a dissected Su-17 wing erected beside it. The aircraft appears to be completely intact.

Su-15 weapons testbed (c/n 0615316)

A Su-15 *sans suffixe* coded '16 Blue' (c/n 0615316) was used by the Sukhoi OKB as a weapons testbed, carrying special equipment pods on the outer wing pylons.

T-58N experimental nuclear strike aircraft

Until the 1970s the Soviet Air Force required virtually all Soviet fighter types to be capable of carrying small-calibre (ie, tactical) nuclear bombs – originally because no supersonic tactical bombers existed and then probably as an insurance policy in case the bombers were destroyed by enemy strikes. Fighters adapted for the 'nuke 'em' role received the suffix letter N to the designation standing for *nositel'* – lit. 'carrier', ie, delivery vehicle for *special stores* (*spetsizdeliya*), as the nuclear munitions were coyly referred to.

The Su-15 also 'fell victim' to this trend. It is known that a single Su-15 was modified as the T-58N tactical nuclear strike aircraft. No details have been disclosed to date.

Su-15UT conversion trainer (U-58T, *izdeliye* 42)

It so happened that dual-control trainer versions of Soviet combat aircraft usually appeared much later than the baseline single-seaters for some reason. The Su-15 was no exception.

The first project studies of a two-seat trainer version of the T-58 dated back to 1961-62 but the design work had to be suspended due to pressure of higher-priority programmes. Hence, in keeping with the well-established tradition, the Su-15 had to be mastered by service units without the benefit of a *sparka*. Officially development of the trainer version was initiated only on 30th April 1965 by a special item of the aforementioned CofM directive clearing the Su-15 for service. The appropriate MAP order appeared on 20th May 1965, requiring OKB-51 to build two prototypes and a static test airframe; all three were to be manufactured by the Novosibirsk aircraft factory and the prototypes were to commence State acceptance trials in the second quarter of 1967. The aircraft received the manufacturer's designation U-58 (ie, *oochebnyy* [*variahnt samolyota T-*] *pyat'desyat vosem'* – trainer version of the T-58).

Design work on the trainer version continued for the greater part of 1965; the in-house project review and the sessions of the mock-up review commission took place in October. Since the U-58 was intended for training pilots both in flying the aircraft and in combat tactics, it was to retain the complete avionics and armament fit of the single-seater and have similar performance.

The trainer differed from the single-seat version in having a second cockpit for the instructor; this required a 450-mm (1 ft 5¾ in) fuselage stretch to be made aft of the existing cockpit and the capacity of the No.1 fuselage fuel tank to be reduced. The cockpits were enclosed by a common canopy similar to that of the Su-9U, with individual aft-hinged portions and a fixed section in between. Each cockpit had a full set of controls and instruments and a KS-4 ejection seat. The canopy of the rear cockpit incorporated a retractable periscope to give the instructor a measure of forward view during take-off and landing. Changes were made to some systems.

The *sparka*'s avionics and equipment were to match that of the late-production (ie, upgraded) Su-15. Thus the trainer was meant to be armed with two R-98 AAMs and equipped with an advanced ***Korshoon***-58 (Kite, the bird) fore control radar, a derivative of the Oryol-D58 radar.

A full set of manufacturing documents for the U-58 was delivered to the Sukhoi OKB's Novosibirsk branch office by September 1966. By then it was obvious that the manufacture of the trainer airframes at the production plant was running behind schedule and the delivery dates (and hence the State acceptance trials deadlines) would have to be shifted. Also, in 1967 the Air Force demanded that the new *Taifoon* (Typhoon) fire control radar be integrated on the upgraded (ie, double-delta) Su-15 instead of the Korshoon radar; this meant the Taifoon would have to be fitted to the trainer as well, causing further delays. Hence the OKB suggested splitting the U-58 programme into two stages to speed up progress – a suggestion gladly accepted by MAP. Stage One involved developing a simplified conversion trainer lacking radar and some other equipment items; these would be integrated during Stage Two to create a fully capable combat trainer.

The downgraded conversion trainer variant received the manufacturer's designation U-58T and the service designation Su-15UT (*oochebno-trenirovochnyy* – for [conversion and proficiency] training) to indicate it had no combat capability. The aircraft featured a standard navigation/communications suite as fitted to the single-seat Su-15 but lacked the latter's fire control radar, Lazoor'-M GCI command link system, radar warning receiver and missile arming/launch system modules. On the other hand, the avionics were augmented by the addition of an SPU-9 intercom and an MS-61 cockpit voice recorder. The bottom line was that the trainer's avionics fit was rather basic. Like the single-seater, the Su-15UT had missile pylons but these could carry only dummy missiles. The addition of a second cockpit reduced the capacity of the No.1 fuselage fuel tank by 900 litres (198 Imp gal); by way of partial compensation a fifth fuel tank holding 180 litres (39.6 Imp gal) was added in the rear fuselage beneath the engines. All of this caused a rearward shift in the CG position, which was restored by installing a counterweight in the forward fuselage. Empty weight increased to 10,660 kg (23,500 lb); this, together with a reduction of the fuel load to 5,100 kg (11,240 lb), entailed a substantial reduction in range.

The static test example of the U-58T was completed in late 1967. Designated U58T-1, the first prototype followed in the summer of 1968; the aircraft had pure delta wings and the BLC system was inactive. A group of Sukhoi OKB specialists headed by engineer in charge L. A. Ryumin arrived from Moscow in early August to take charge of the aircraft. Ye. K. Kukushev was appointed project test pilot, performing the U58T-1's maiden flight on 26th August; sometime before 16th September he flew the aircraft to Zhukovskiy. The OKB was unable to hold a full-scale manufacturer's test programme there because MAP kept urging that State acceptance trials begin as soon as possible; thus, as early as 2nd October the first prototype was ferried to the GK NII VVS base at Vladimirovka AB. Manufacturer's flight tests continued there, proceeding in parallel with the State acceptance trials and ending on 12th December 1968.

On 15th-19th October 1968 the U58T-1 was test flown by a number of service pilots from operational PVO units, including the new IA PVO C-in-C Lt. Gen. A. L. Kadomtsev. The actual flights under the State acceptance trials programme did not begin until 16th November. GK NII VVS project test pilots M. I. Bobrovitskiy, G. A. Bayevskiy and N. V. Rukhlyadko did the testing, additional test flights being made by Stepan A. Mikoyan, A. A. Manucharov, V. S. Kotlov and S. A. Mayorov. The programme included performance testing, assessment of field performance, stability and handling; part of it was performed by OKB test pilots, as was customary.

One of the flights on 20th January 1969 nearly ended in an accident; after a mission involving a series of high-g manoeuvres OKB test pilot Ye. K. Kukushev discovered that the main landing gear units would not extend. The pilot again resorted to vigorous manoeuvring and the g force wrenched the gear loose on the fifth try, after which the aircraft landed normally. It turned out that the gaps between the main gear doors and the wheel well walls were too small; this had not caused any trouble until then, the defect surfacing only when the aircraft was subjected to high g loads.

The State acceptance trials were completed on 26th February 1969. Predictably, the two-seater's performance was inferior to that of the single-seater: range had dropped to 1,390 km (863 miles) and the service ceiling had decreased to 17,700 m (58,070 ft). The stability and handling were deemed acceptable, except that some directional instability set in above Mach 1, which was a serious drawback. Generally, however, GK NII VVS stated that the Su-15UT was suitable as a pilot trainer, except for weapons training which was beyond the aircraft's capabilities.

The final protocol of the State commission was signed in the summer of 1969, whereupon the OKB set to work eliminating the deficiencies discovered in the course of the trials. These included the instability mentioned above, and the cause was clear – the vertical tail area was too small, now that the fuselage was stretched and the area ahead of the CG had increased. Increasing the vertical tail area by inserting another 'plug' at the root, as had been the case in the Su-15's early flight test days, was impossible for structural strength reasons. The OKB decided to try equipping the Su-15UT with ventral fins. In the spring of 1970 OKB test pilots Yevgeniy S. Solov'yov and A. N. Isakov performed a special test programme in the suitably modified U58T-1, comparing the aircraft's handling with and without ventral fins. It turned out that the fins did not give any major improvement; therefore the OKB and the Air Force compromised, limiting the trainer's maximum permissible speed to Mach 1.75. Consequently the service ceiling further decreased to 16,700 m (54,790 ft), as the above figure of 17,700 m had been obtained at Mach 1.9, using the standard climb technique.

The Su-15UT entered full-scale production in Novosibirsk in 1969 under the in-house product code *izdeliye* 42. The first production aircraft (c/n 0115301, often shortened to 0101) was completed in October, making its first flight on 10th December with factory test pilots V. T. Vylomov and V. A. Belyanin at the controls. Production examples differed from the prototype in having double-delta wings and an operational boundary layer control system. The first five Su-15UTs were delivered to operational units in the spring of 1970; in July that year the aircraft was officially included into the inventory pursuant to Ministry of Defence order No.0115.

Shortly afterwards the Su-15UT passed a separate spinning test programme. Since the trainer's directional stability parameters were different from those of the single-seater, it was though that the spinning characteristics might be different, too.

The Su-15UT remained in production until the end of 1972. The final batches featured a new R-832M Evkalipt-SMU communications radio replacing the earlier R-802V. The empty operating weight of the production model grew to 10,750 kg (23,700 lb).

In addition to first-line units, a number of Su-15UTs was delivered to the PVO's Stavropol' Military Pilot College (VVAUL – *Vyssheye voyennoye aviatsionnoye oochilischche lyotchikov*) and the Test Pilots School in Zhukovskiy (ShLI – *Shkola lyotchikov-ispytahteley*), the Soviet/Russian counterpart of the ETPS. Unfortunately on 17th April 1987 a Su-15UT operated by ShLI suffered a critical failure; the crew ejected but cadet A. V. Chechoolin lost his life.

The Su-15UT had the NATO reporting name *Flagon-C*.

U-58B combat trainer prototype (*izdeliye* 37UB)

Completed much later than intended due to late deliveries of the mission avionics, a single Su-15 was built as a fully capable combat trainer designated U-58B (*boyevoy* – combat, used attributively). Known at the Novosibirsk aircraft factory as *izdeliye* 37UB (somewhat surprisingly, the product code was derived from that of the single-seater, not the Su-15UT), the aircraft with the tactical code '70' and the non-standard c/n 0003УБ86 (ie, 0003UB86) entered flight test in the summer of 1970. On 24th June the U-58B made its first flight at the hands of factory test pilot A. S. Gribachov and was ferried to the OKB's flight test facility in Zhukovskiy on 2nd August. Originally the aircraft was powered by R11F2S-300 engines but these were replaced by R11F2SU-300s in May 1971, allowing the BLC system to be activated.

Outwardly the U-58B was identifiable by the reinforced twin-wheel nose gear unit identical to that of the T-58L development aircraft. Unlike the latter, however, the purpose of this modification was to cater for the higher weight of the forward fuselage caused by the installation of the radar. The main gear units featured KT-117 wheels and the wings were taken from the T-58N development aircraft which was cannibalised after completing its flight test programme.

As already mentioned, the Su-15UT was 'tail heavy' requiring the provision of ballast in the nose. The U-58B, on the contrary, was 'nose heavy' due to the combination of a rear cockpit and a fire control radar, with the CG located well forward; hence it was as sluggish as a wet sponge and generally disappointing in its performance. At the initiative of the OKB, with MAP's formal agreement, the U-58B's development programme was suspended; a while later the military also gave their consent to this decision. After sitting in storage at the OKB's flight test facility for a while the aircraft was transferred to one of the Air Force's technical schools as a ground instructional airframe.

Su-15UT research aircraft for exploring weather minima

In 1973 LII used a suitably modified Su-15UT (c/n 0815315) to conduct research for the purpose of determining the weather minima in which aircraft could take off and land safely. For example, the Su-15's weather minima were 250 x 2,000 m (ie, cloudbase 250 m/820 ft and horizontal visibility 2,000 m/1.24 miles) in the daytime and 350 x 3,000 m (1,150 ft x 1.86 miles) at night; for the Su-15TM interceptor equipped with the RSBN-5S short-range radio navigation system (and described later in this chapter) the minima were 60 x 800 m (200 ft x 0.5 mile).

Tragically, one of the test missions on 29th October 1975 aimed at reducing the type's weather minima ended in a crash in which LII test pilot Ivan V. Makedon and 148th

The crew of a Su-15UT coded '95 Blue' climbs out during a friendly visit by an IA PVO unit to Poland. Note the blind flying hood in the forward cockpit and the forward vision periscope in the rear cockpit.

TsBP i PLS pilot Rudol'f N. Zoobov were fatally injured. Landing with the blind flying hood closed and a simulated control system actuator failure, Zoobov (who was the pilot in command) inadvertently increased the sink rate from the required 5-6 m/sec (980-1,180 ft/min) to 10 m/sec (1,970 ft/min) and then to 15 m/sec (2,950 ft/min). Despite corrective action the aircraft landed hard between the outer and inner marker beacons, undershooting by 1,400 m (4,590 ft); next thing, it collided with the elevated edge of a concrete road and caught fire, both pilots being thrown clear of the cockpit before the machine came to a halt. They were rushed to a hospital but Zoobov was dead on arrival, Makedon succumbing to his injuries on 23rd November.

Su-15T interceptor (T-58T, Su-15 Stage II upgrade, *izdeliye* 37M or *izdeliye* 37 Srs M, *izdeliye* 38)

As already noted, the performance of the production Su-15 did not satisfy the military completely; the type was due for a mid-life update. Such an upgrade was developed in due course, and in contemporary documents the aircraft thus upgraded was referred to as the *Su-pyatnadtsat' vtorovo etapa* ('Su-15 Stage II'). The requirements for this version were outlined by the aforementioned CofM directive of 30th April 1965. At that time the Sukhoi OKB was swamped with work, being busy with such programmes as the T-4 (*izdeliye* 100) heavy supersonic missile strike aircraft, the S-22I 'swing-wing' fighter-bomber (the Su-17 prototype), the T-58M attack aircraft project (soon to be redesignated T-6) and so on, so it was less than overjoyed at the prospect of having to upgrade the Su-15.

Most of the design work on the Su-15 upgrade proceeded on a 'time permitting' basis due to the programme's low-priority status. Actually the work did not begin in earnest until early 1966, and even then for the first two years it made painfully slow progress. The reason was that the type of radar to be fitted was not decided on for a long time.

In mid-1966 the OKB started detail design work on incorporating the new advanced Korshoon-58 fire control radar, the SAU-58 AFCS, the RSBN-5S Iskra (Spark) short-range radio navigation system (RSBN = *rahdiotekhnicheskaya sistema blizhney navigahtsii* – SHORAN), a new communications radio and a skid landing gear on the Su-15. Later the military belatedly discovered that the Korshoon radar could not provide the required performance; the attention now focused on the Smerch radar designed for installation in the Mikoyan Ye-155P heavy interceptor (the future MiG-25P). As a result, in October 1967, when the OKB had all but completed the project documents for the installation of new systems and equipment on

The U-58B combat trainer version of the Su-15UT ('70', c/n 0003UB86) seen during a test flight. The aircraft displayed disappointing performance and remained a one-off.

the Su-15, it was greeted by a joint ruling by several ministries to the effect that all work on the Korshoon-58 radar be stopped and the fighter be equipped with the Taifoon fire control radar (a derivative of the Smerch) instead. Officially this change was documented by a ruling of the Military Industry Commission on 22nd March 1968. Thanks a bunch. It was back to the good old drawing board.

The upgrade programme was divided into two stages. Stage One involved State acceptance trials of the aircraft with the existing R-98 AAMs in November 1968; during the second stage scheduled for the third quarter of 1969 the aircraft was to be armed with the new K-98M missile (which had not yet passed its State acceptance trials either). The aircraft modified to Stage One specifications was to be designated Su-15T, the suffix letter referring to the Taifoon radar, while the Stage Two aircraft would be designated Su-15TM. This gave the avionics and defence industry a respite, allowing the aircraft's new avionics and armament to be put through their paces.

The Sukhoi OKB had no choice but to rework the project documents all over again. The advanced development project of the Su-15T interceptor (known in-house as the T-58T) was completed by early September 1968, the in-house project review and the sessions of the mock-up review commission taking place in October. The aircraft was to be powered by the new and more powerful Gavrilov R13-300 engines; changes to the hydraulics and electrics were also envisaged.

The military demanded that both the Su-15T and the Su-15TM should have a secondary strike capability that would make them usable by the Soviet Air Force's tactical arm (FA – *Frontovaya aviahtsiya*). The weapons options in strike configuration were one or two 500-kg (1,102-lb) bombs; up to four 100- or 250-kg (220- or 551-lb) bombs; one or two UB-16-57U rocket pods (*oonifitseerovannyy blok* – standardised [rocked] pod), each holding 16 57-mm (2.24-in) S-5 folding-fin aircraft rockets; one or two S-24 heavy unguided rockets; and two UPK-23-250 cannon pods (these were to be carried under the wings, not on the fuselage hardpoints). The possibility of incorporating an internal cannon was also considered.

The control system was also to be modified by incorporating the servo drives of the SAU-58 AFCS which was finally nearing the end of its development at the Sukhoi OKB. Apart from the usual speed/altitude/angle stabilisation and 'panic button' (automatic restoration of straight and level flight) functions typical of any automatic flight control system, the SAU-58 was to enable automatic flight along several preset trajectories and automate the main stages of the interception process. Additionally, the AFCS was to enable automatic low-level terrain-following flight. This totally new feature was not meant for air defence penetration, of course; rather it resulted from the customer's new stringent requirement that the upgraded Su-15 was to be capable of intercepting targets flying at altitudes right down to 500 m (1,640 ft). Since the Taifoon radar still lacked 'look-down/shoot-down' capability, the intention was to 'paint' the targets from below, flying at altitudes less than 500 m. This meant flying dangerously close to the terrain at speeds close to 1,000 km/h (620 mph), and the pilot had to work the radar all the while! Hence the OKB opted for a simpler version of the AFCS relying on the radio altimeter, not the radar, as the primary source of data; this made flight level stabilisation possible but the feature was usable only over flat or very moderately hilly terrain.

The 'Su-15 Stage II' was to feature a new R-832M Evkalipt-SM communications radio replacing the earlier R-802V, a Pion-GT (Peony, pronounced *pee on*… sorry) antenna/feeder system combining the receiver and transmitter antennas of various avionics into a single ensemble for greater reliability, built-in test equipment (BITE) and an RSBN-5S Iskra-K SHORAN. The latter enabled semi-automatic landing approach down to 50-60 m (165-200 ft), improving the aircraft's all-weather capability considerably.

An early-production (pure-delta) Su-15 (c/n 0515348) was converted as the Su-15T prototype; for security reasons this aircraft was referred to at the OKB as 'aircraft 0006'. With the Air Force's consent the prototype entered test with an incomplete avionics fit, lacking the BITE and SHORAN, among other things; nor were the R13-300 engines (the prototype was powered by R11F2S-300s), skid landing gear and strike armament installed. On the other hand, the aircraft was rewinged, receiving double-delta wings; a prototype Taifoon radar, the SAU-58 AFCS and a twin-wheel nose gear unit were installed.

On the night of 4th January 1969 the modified aircraft was delivered to the flight test facility in Zhukovskiy. V. A. Krechetov was appointed project test pilot for the manufacturer's test phase, with M. L. Belen'kiy as engineer in charge. According to some documents, the Su-15T made its maiden flight on 27th January; however, Krechetov's log book says 31st January. As the military kept pushing the Sukhoi OKB to submit the fighter for State acceptance trials, the manufacturer's tests were suspended after only eight flights and on 6th March 1969 Krechetov flew the Su-15T prototype to GK NII VVS.

The acceptance procedure dragged on until the end of May; meanwhile the OKB carried on with the manufacturer's flight tests. IA PVO Vice-Commander Col. Gen. F. Smetanin was chosen to chair the State commission, while R. N. Lazarev headed the Air Force's flight test team. The latter included GK NII VVS test pilots S. I. Lavrent'yev, Eduard M. Kolkov, V. I. Petrov and Stepan A. Mikoyan, as well as OKB test pilot V. A. Krechetov. The trials proceeded in two stages, the first of which was to be completed in the first quarter of 1970.

The State acceptance trials of the Su-15T prototype were a far cry from those of the original T-58D, proceeding slowly and laboriously. The main objective of the trials was to assess the interceptor's combat capabilities with the new radar. Here there would appear to be no pitfalls, since fire control radars broadly similar to the Taifoon had been verified on the production Tu-128 heavy interceptor and the Ye-155P which had passed its trials successfully. Nevertheless the Taifoon radar turned out to be rather troublesome. By the end of the year the prototype had made 64 flights but only 40 of them counted, the other 24 missions apparently being aborted due to malfunctions (or being training flights and the like). To avoid delaying the Su-15T's production entry GK NII VVS decided to issue a so-called preliminary report; in February 1970 Stage A of the trials was discontinued altogether. The report said that only 63 of the 87 flights made had proceeded in accordance with the plan of the trials, the remainder being devoted to perfecting the radar and the AFCS. Eventually the OKB managed to get the interceptor's principal systems up to scratch and the aircraft was ready for Stage B of the trials.

Here we have to go back a bit. In December 1969 the Sukhoi OKB completed the second T-58T prototype converted from the fifth production Su-15 (c/n 0115305); A. Sholosh was appointed engineer in charge of this aircraft's test programme. This aircraft was actually not a Su-15T but the Su-15TM prototype and featured an almost complete avionics and equipment fit, including R13-300 engines and an upgraded Taifoon-M radar compatible with K-98M missiles instead of the basic Taifoon. Since the State acceptance trials of the first prototype Su-15T were running late, it was decided that the second prototype should join the programme at this stage. Due to development problems with a number of systems it was more than three months before the second prototype could enter flight test; the aircraft first flew on 7th April 1970 with V. A. Krechetov at the controls. Four days later it was ferried to GK NII VVS at Vladimirovka AB. By then the institute had been conducting live weapons tests and combat tactics verification with the first aircraft for more than three weeks in accordance with Stage B of the trials. The availability of the second prototype powered by the intended engines made it possible to determine the aircraft's flight performance.

The State acceptance trials ended in mid-June 1970; Stage B was not completed in full either, as the MAP and Air Force top brass demanded the beginning of the Su-15TM's State acceptance trials pronto (this was exactly what the second prototype was intended for). In the course of Stage B the two aircraft made 58 and 14 flights respectively. The final report of the trials said that generally the weapons system met the specifications, even though the new equipment, first and foremost the Taifoon radar (no pun intended), proved rather unreliable.

The Novosibirsk aircraft factory was to complete the first 20 production Su-15Ts in the second half of 1969; yet the first production example of the new model (referred to initially as *izdeliye* 37M or *izdeliye* 37 Srs M) was flown by the factory's CTP V. T. Vylomov only on 20th December 1970. As the trials of the more advanced Su-15TM progressed, the interest of the military in the interim Su-15T waned, and the production run was limited to 20 aircraft (c/ns 0115301 through 0115310 and 0215301 through 0215310). Confusingly, the new version had a separate batch numbering sequence; once again, only the batch number and the number of the aircraft in the batch were painted on. The c/ns are sometimes quoted with 06 or 07 added at the front (eg, 070215301); the meaning of these digits is unknown. Later, when the Su-15TM (likewise designated *izdeliye* 37M) entered production, the Su-15T's product code was changed to *izdeliye* 38 to discern it from the newer model.

The avionics suite of the production Su-15T included a Taifoon radar, an SAU-58 AFCS, an R-832M Evkalipt-SM radio, an

Top and above: '37 Red', the second prototype Su-15T (c/n 0115305) – or rather the Su-15TM prototype – at the PVO's technical school in Solntsevo. The aircraft later moved to the museum at Moscow-Khodynka.

RSBN-5S Iskra-K SHORAN, an ARL-SM (Lazoor'-SM) GCI command link receiver, an ARK-10 ADF, an RV-5 low-range radio altimeter, an SPO-10 Sirena-3 radar warning receiver and a Pion-GT antenna/feeder system. The latter's antennas were mounted on the air data boom at the tip of the radome and in a small fairing above the brake parachute container (which required the rudder to be cut away at the base). The nose gear unit featured twin KN-9 non-braking wheels. The production Su-15T was powered by the old R11F2SU-300 engines.

On 25th January 1971 GK NII VVS commenced checkout trials of the first and 15th production Su-15Ts (the c/ns are stated in abbreviated form as 0101 and 0205) with the objective of checking the operation of the armament. The former aircraft made only three flights under this programme and the other machine 23. The programme was completed within a month, and the results were disheartening: the aircraft proved incapable of intercepting low-flying targets. In five test flights at an altitude of 500 m (1,640 ft) the radar tracked the target on two occasions only, and then at very close range – a matter of 3 km (1.86 miles), which was totally unsatisfactory. Worse, neither could the radar reliably capture and track the target at higher altitudes almost throughout the permitted missile launch range. Besides, poor electromagnetic compatibility (EMC) of various avionics was discovered, landing with the BLC system switched on was complicated due to the limited tailplane travel and so on.

As a result, the delivery of the production Su-15Ts dragged on for more than 12 months due to the need to rectify these defects. The aircraft did not become operational until the summer of 1972. Most of them were eventually transferred from active duty to the Stavropol' VVAUL with the Air Force's and MAP's approval, serving on as trainers. A few were transferred to MAP for use in various R&D programmes.

The Su-15T's NATO reporting name was *Flagon-E*.

Su-15T multi-role testbed (c/n 0206)

The 16th production Su-15T ('short' c/n 0206) was one of the abovementioned research vehicles. In 1972 the Sukhoi OKB and LII held a joint programme with this aircraft to explore the EMC parameters of the avionics and test a trim mechanism in the roll control circuit. At the end of the year the aircraft was fitted with increased-area tailplanes, undergoing a special test programme in 1972-73 with a view to improving the handling in landing mode with the BLC activated. In 1973 Su-15T c/n 0206 served as a testbed for the new Tester-U3 FDR which later became standard on many Soviet combat aircraft.

Buddies: the Sakhalin experience

The Sukhoi OKB's first involvement with in-flight refuelling (IFR) dates back to 1971. The need to master IFR techniques was caused by the necessity to increase the range and combat radius of the Su-24 tactical bomber which was undergoing trials at the time. By the end of the year the OKB prepared a set of project documents envisaging the installation of an IFR system on what was then the T-58M.

Unlike the Sukhoi OKB, which was breaking new ground with in-flight refuelling, the Flight Research Institute and the Zvezda (Star; pronounced *zvezdah*) design bureau (formerly OKB-918) under Guy Il'yich Severin were old hands at this technique. By the early 1970s these two establishments had accumulated a wealth of experience in developing, testing and using various IFR systems. In the 1950s and 1960s such systems were created in the Soviet Union both for heavy bombers and for tactical aircraft, although it would be a while before the latter would actually benefit from in-flight refuelling.

The Zvezda OKB had been working on IFR systems since the late 1960s and the programme was code-named Sakhalin after an island in the Soviet Far East. (Perhaps the implication was that 'with this system our aircraft will be able to reach Sakhalin non-stop'!) In 1971-73 the OKB's specialists designed and tested the principal components of a hose-and-drogue refuelling system.

The preliminary design project of the T-58M attack aircraft-turned-tactical bomber featuring an IFR system with a retractable refuelling probe was assessed at a session of MAP's Scientific & Technical Council. In keeping with the Council's recommendations the Sukhoi OKB decided to hold a series of tests in advance so that the system would be fully mastered by the time the bomber was ready to take it. The Su-15 was selected for conversion into IFR system testbeds, one aircraft acting as the tanker and the other as the receiver. In October 1973 the VVS issued a specification in which three Sukhoi types – the Su-15TM, the Su-17M/Su-17M2 and the T-6 (by then allocated the service designation Su-24) – were stated as possible tanker aircraft fitted with 'buddy' refuelling pods. The pod itself, or hose drum unit (HDU), was designated UPAZ (*oonifitseerovannyy podvesnoy agregaht zaprahvki* – 'standardised suspended (ie, external) refuelling unit').

Pursuant to a joint MAP/Air Force ruling the Sukhoi OKB allocated two Su-15s – the first pre-production aircraft (c/n 0015301) and an early production aircraft (c/n 0215306) – for testing the Sakhalin-1A IFR system. Both aircraft by then had a long history as 'dogships', having been used for testing various systems and equipment. Additionally, LII allocated one of its Su-15s ('37', c/n 1115337) for testing the Sakhalin-1A system and evolving IFR techniques. Since each of these aircraft served a different purpose, they will be described separately hereunder, followed by an account of the trials.

Su-15 'buddy' IFR tanker testbeds

By December 1972 Su-15 '34 Red' (c/n 0015301) had been modified as the first of two examples fitted with a dummy UPAZ-1A (Sakhalin-1A) pod for aerodynamic testing (actually the pod featured a functional hose drum allowing the drogue to be deployed and retracted). On 19th December OKB test pilot V. A. Krechetov performed the first test flight in order to assess the aircraft's stability and handling with this rather bulky external store.

In 1973 the OKB issued a set of documents for the adaptation of the production-standard Su-15's fuel system to accept the UPAZ pod. This involved installation of additional fuel pumps in the Nos 1, 2 and 3 fuselage tanks, modifications to the electric system and replacement of the standard TRV1-1A fuel metering kit with a new TRK1-1 kit. The display of the Oryol fire control radar was removed to make room for the HDU's control panel.

The UPAZ-1A had a 26-m (85 ft 3 in) hose and a flexible 'basket' drogue. The hose drum was powered by a ram air turbine (RAT) with an intake scoop on the port side which was normally closed flush with the skin. A second air intake at the front closed by a movable cone was for an RAT driving a generator for the electric transfer pump. Normal delivery rate was 1,000 litres (220 Imp gal) per minute but this could be increased to 2,200 litres (484 Imp gal) in case of need. 'Traffic lights' were installed at the rear end of the pod to indicate fuel transfer status to the pilot of the receiver aircraft. The pod was attached to the airframe via a so-called standardised piping/electrics connector (UURK – *oonifitseerovannyy oozel razyoma kommunikatsiy*) and could be jettisoned pyrotechnically in an emergency. A test equipment suite was fitted.

Outwardly the aircraft, which received a new tactical code ('01 Red') after the conversion, differed from its standard sister ships in having an egg-shaped pod housing an AKS-5 camera fitted to the starboard PU-1-8 missile launch rail. The aft-looking camera served for filming the moments when the receiver aircraft made and broke contact.

The conversion was performed in April-July 1974 by the OKB's Novosibirsk branch. On 4th July 1974 OKB test pilot A. S. Komarov made the first checkout flight from Novosibirsk-Yel'tsovka, whereupon the modified aircraft was flown back to Zhukovskiy for testing.

The second aircraft to be fitted with a dummy UPAZ-1A pod was the aforementioned Su-15 coded '37' (c/n 1115337). Manufactured

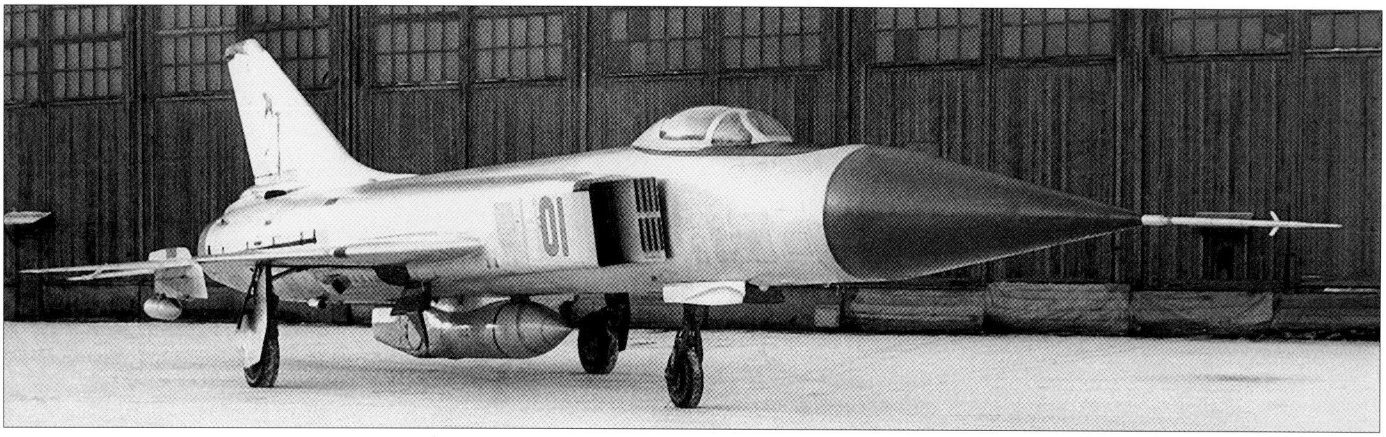
01

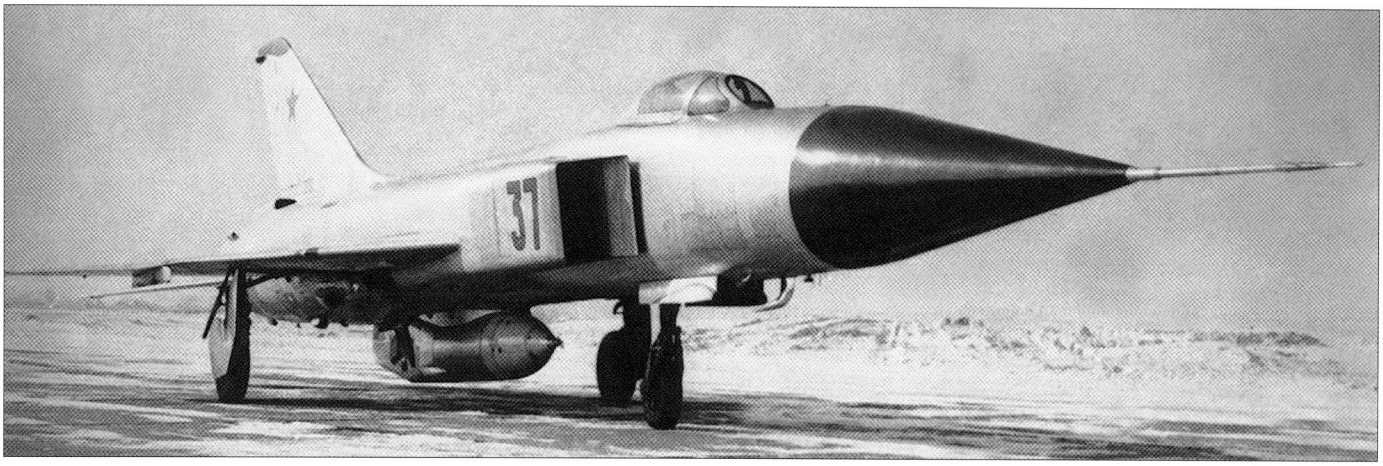
37

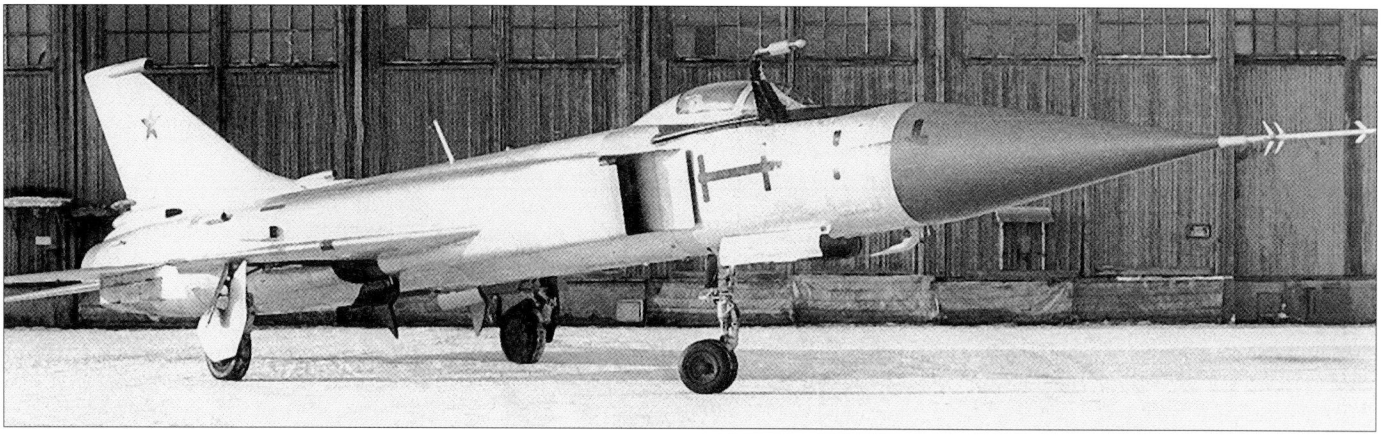

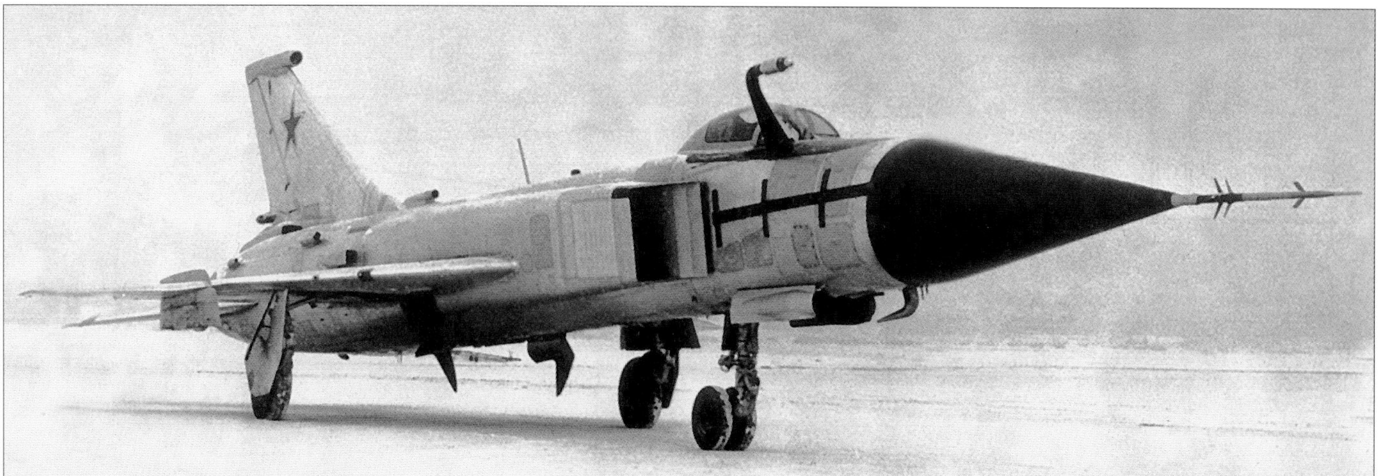

on 13th February 1970 and delivered new to LII for special tests, this aircraft featured double-delta wings and an operational BLC system; also, it could be fitted with both R11F2SU-300 engines and R13-300s. In late 1973 after completing a spinning test programme the fighter was modified in house by LII for installation of an operational UPAZ-1A. Su-15 c/n 1115337 was intended for testing the pod proper and for verifying the approach and contact technique. Contact would be made in 'dry' mode (ie, without actual fuel transfer), hence the scope of the modification work on this aircraft was much smaller as compared to Su-15 c/n 0015301. The conversion involved removing the radar, which was replaced by ballast, and installing the HDU control panel, the UURK 'wet' centreline pylon and the appropriate data recording equipment. The missile launch rails were removed for the duration of the Sakhalin test programme.

In 1974 LII test pilots held a series of tests on this aircraft, checking the operation of the HDU (ie, hose deployment/retraction and the operation of the pod's other systems). Afterwards this aircraft continued in use with the institute, participating in several research programmes. Unfortunately on 24th December 1976 it crashed near Lookhovitsy, Moscow Region, killing LII test pilot Leonid D. Rybikov. The exact cause was never determined. Five

Opposite page, top to bottom:

'01 Red' (formerly '34 Red', c/n 0015301), the first Su-15 converted into a 'buddy' tanker equipped with a UPAZ-1A (Sakhalin-1A) hose drum unit. The large size of the pod is readily apparent. Note the photo calibration markings on the rear fuselage and tail and the AKS-5 camera pod on the starboard wing pylon.

'37 Red' (c/n 1115337), the second Su-15 modified to take the UPAZ-1A HDU. Note the absence of the missile launch rails.

Two views of Su-15T c/n 0206 following conversion as the receiver aircraft for the Sakhalin test programme with a fixed IFR probe. Note the camera fairing on top of the fin, the photo calibration markings and the non-standard comms radio aerial on the centre fuselage.

This page, top to bottom:

The Su-15T receiver aircraft makes contact with Su-15 '01 Red'. The tanker's flaps are deployed and the receiver aircraft's airbrakes are also deployed for station-keeping. Note the white stripes on the hose showing how much of it has been paid out.

An air-to-air of '37 Red' as it streams the drogue.

The receiver aircraft has made stable contact with the drogue; fuel transfer can begin now.

This picture taken by the camera on the receiver aircraft's tail illustrates the offset of the IFR probe and the design of the drogue.

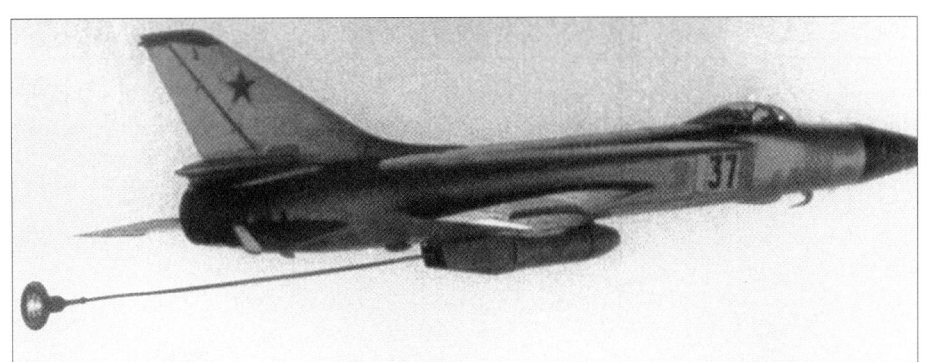

(some sources say three) minutes after taking off from Tret'yakovo airfield (the factory airfield in Lookhovitsy) on a positioning flight to Zhukovskiy, which is about 90 km (56 miles) away as the crow flies, the aircraft inexplicably entered a shallow dive with considerable bank, coming down in a field 39.5 km (24.5 miles) from the point of origin and disintegrating utterly. Examination of the crash site and analysis of the wreckage showed that the aircraft had been perfectly serviceable up to the point of impact. At the final section of the flight path the aircraft had been levelled out but continued losing altitude due to the high sink rate prior to that, impacting in a wings-level attitude; the pilot did not eject and was killed. Temporary pilot incapacitation was eventually cited as the cause of the crash.

Shortly after the completion of the Sakhalin IFR system's trials Su-15 c/n 0015301 was returned to the PVO as time-expired. Actually the aircraft could have been returned to service after an overhaul and a service life extension, but in view of the extensive conversion it had undergone, using it as a combat aircraft was inadvisable. Hence '01 Red' was transferred to a PVO Junior Aviation Specialists School (ShMAS – *Shkola mlahdshikh aviatsionnykh spetsialistov*) located in Solntsevo just outside the Moscow city limits as a ground instructional airframe. After the closure of the school in 1991 the Su-15 was returned to the OKB for preservation and, after cosmetic repairs, joined the collection of the open-air aviation museum at Moscow-Khodynka established in 1994. Regrettably, as already mentioned, the museum is now closed and the exhibits are in serious danger.

Su-15T IFR system testbed (receiver aircraft)

In late 1973 the Sukhoi OKB started converting the abovementioned Su-15T c/n 0206, another long-serving 'dogship', into the receiver aircraft under the Sakhalin programme. Again the conversion was quite extensive, as the aircraft had to be equipped with a completely new IFR system ensuring the correct fuel transfer sequence in order to keep the CG within the allowed limits.

The aircraft was equipped with a new TRK1-1 fuel metering kit and a fixed L-shaped refuelling probe offset to starboard ahead of the cockpit windscreen; the tip of the probe was located 4.25 m (13 ft 11⅜ in) aft of the tip of the radome, 0.9 m (2 ft 11½ in) to the right of the fuselage centreline and 0.8 m (2 ft 7½ in) from the fuselage surface. A forward-looking AKS-5 camera was installed in a cigar-shaped fairing replacing the dielectric fin cap of the communications radio antenna for filming the contact with the tanker's drogue; hence a non-standard ASM-1 blade aerial serving the R-832M radio was fitted to the centre fuselage. An MSL-2 (aka KhS-62) flashing beacon was installed on the upper centre fuselage for synchronising the operation of all photo and movie cameras capturing the refuelling sequence. Since the refuelling operation imposed considerable stress on the pilot of the receiver aircraft, Su-15T c/n 0206 was fitted with special *Koovshinka* (Water lily) medical equipment recording the pilot's physiological parameters (pulse and breathing) and recording the distribution of his concentration during various stages of the process. Finally, a PAU-467 cine camera was fitted.

The course of the Sakhalin programme

Apart from the three aircraft described above, the tests involved a MiG-21U owned by LII which acted as a chase plane and camera ship. According to plan the first stage of the tests (performed jointly by the Sukhoi OKB and LII) was devoted to evolving and perfecting the optimum approach and contact technique. Yevgeniy S. Solov'yov was the OKB's project test pilot, with M. L. Belen'kiy as engineer in charge; LII appointed V. D. Koorbesov and Yu. N. Goonin respectively.

Before the joint flights and attempted contacts with the tanker could begin, the probe-equipped Su-15T made a series of test flights to see if and how the refuelling probe affected the aircraft's stability and handling. The actual flight test programme commenced on 31st May 1974 when the first two flights were made. The approach to the tanker and attempted contacts were made at 8,200 m (26,900 ft) on the first occasion and about 6,000 m (19,680 ft) on the second occasion; the speed was 550 km/h (340 mph) in both cases. Both attempts ended in failure; moreover, on the second try the fuel transfer hose got entangled with the refuelling probe and snapped as the receiver aircraft manoeuvred.

A pause was then called while the engineers made corrections to the piloting technique during the final approach phase. The pause turned out to be a long one, the flights resuming only on 24th December 1974. During this period the Zvezda OKB revised the UPAZ-1A pod, increasing the length of the hose to 27.5 m (90 ft 2 in). In the meantime Yevgeniy S. Solov'yov took special training on a purpose-built simulator at the OKB and in the actual aircraft (the approach to the tanker was simulated on the ground and special reference lines were applied to the cockpit canopy to mark the drogue's position at different ranges). It was decided that the tanker aircraft would deploy the flaps and that the dampers in the receiver aircraft's control circuits would be switched on. The flight altitude was reduced to 4,000 m (13,120 ft) and the tanker approach technique was revised accordingly.

The effect of the additional training taken by the pilots was felt immediately; as early as 14th January the modified Su-15T made the first stable contact with the tanker, maintaining refuelling formation for a while. This flight marked the end of the phase involving LII's 'buddy tanker' (Su-15 c/n 1115337), as the OKB's own 'buddy tanker' (Su-15 c/n 0015301) had been completed by then.

Solov'yov remained the receiver aircraft's project test pilot at this stage; the other aircraft was flown by A. S. Komarov and V. A. Krechetov. M. L. Belen'kiy likewise continued in his capacity as engineer in charge of the tests.

Duly prepared and endorsed at all levels, the test schedule for the summer of 1974 envisioned 70 flights, 29 of which would be made by the two aircraft together. Yet the beginning of the programme's main phase involving contacts with the tanker was delayed until 21st January 1975. The first contacts were made in 'dry' mode (without fuel transfer). The missions were flown at altitudes of 2,000-7,500 m (6,560-24,600 ft) and speeds of 480-660 km/h (300-410 mph); each mission involved two to seven attempts at making contact. This stage continued until the end of February, after which another lengthy hiatus followed due to the need to analyse the results obtained and prepare for the next phase. This was devoted to testing and perfecting the UPAZ-1A HDU proper and the associated equipment of the carrier aircraft during refuelling missions.

The work resumed in June 1975. Again, initially contacts were made in 'dry' mode; in early July, however, the plug closing the orifice of the HDU's drogue was removed, allowing fuel to be transferred. An immediate problem arose: on 2nd, 4th and 18th July fuel was seen to be leaking from the drogue after the receiver aircraft's probe had locked into position. It turned out that the drogue had been manufactured inaccurately, making it impossible to adequately seal the joint; a new, properly made drogue took care of the problem. The first successful transfer of 250 kg (550 lb) of fuel took place on 30th July 1975; the following day Su-15T c/n 0206 made two more fuel top-ups, receiving 500 and 1,000 kg (550 and 1,100 lb) of fuel respectively.

By early December 1975 the Sukhoi OKB and the Zvezda design bureau had ironed out all the bugs discovered in the course of the preceding tests. After that, another four flights involving three fuel top-ups were made between 10th and 23rd December; on the latter date the receiver aircraft 'hit the tanker' twice in a single mission, taking on 400 and 250 kg (880 and 550 lb) of fuel respectively. Thus the test programme was successfully completed in full.

The test report said, '...*The in-flight refuelling system and technique verified [...] can be recommended for use primarily on the Su-24 aircraft'*. Further tests of the Sakhalin-1A IFR system were performed on Su-24 bombers.

L.10-10 weapons testbed

In 1978 Su-15T c/n 0206 was modified yet again, becoming a weapons testbed as part of the T-10 fighter's development programme. Designated L.10-10 (ie, 'flying laboratory' No.10 under the T-10 programme?), the aircraft had the standard PU-1-8 missile launch rails replaced with APU-470 launch rails developed for the new K-27E advanced medium-range AAM. The actual test launches of these missiles took place in 1979, the missile eventually entering production as the R-27.

Su-15TM interceptor (T-58TM, *izdeliye* 37M or *izdeliye* 37 Srs M)

As already mentioned, the Su-15TM version of the 'Su-15 Stage II upgrade' commenced State acceptance trials in September 1970 when the prototype (c/n 0115305) joined the programme. The aircraft was flown by GK NII VVS pilots Eduard M. Kolkov, V. V. Migoonov, V. I. Mostovoy and S. I. Lavrent'yev; R. N. Lazarev retained his responsibilities as engineer in charge.

Stage A comprising 40 flights was to verify the operation of the aircraft's principal systems. Progress was terribly slow, only six flights 'for the record' being made by the end of 1970. On 3rd February 1971 the original Su-15T prototype, or 'aircraft 0006' (c/n 0515348), which had been refitted with an upgraded Taifoon-M radar in the meantime, was added to the first aircraft in the hope of speeding up the trials, but little good did it do. During the first three months of the year the two aircraft made a mere 42 flights between them, only three of these 42 missions being accepted 'for the record'. Neither aircraft had the widened air intakes envisaged for the production version which, as noted earlier, had been tested successfully on Su-15 c/n 1315340 ('aircraft 0009'). A third aircraft (c/n 0115309) joined the trials programme in 1971, making 19 test flights with the objective of exploring the structural strength limits.

Meanwhile, plant No.153 was already preparing to build the Su-15T; therefore, as already mentioned, GK NII VVS decided to issue a so-called preliminary report on the trials results in order to avoid holding up production of the modernised fighter. To this end a special test schedule was drawn up, flights under this programme commencing on 20th May 1971. By the end of June the prototypes had made a total of 123 flights under Stage A of the State acceptance trials (including 35 'for the record'). Still, radar operation and the guidance of the radar-homing AAMs at low altitudes was unstable, and the Air Force called a halt to the testing of 'aircraft 0006' for the time being. The other prototype kept flying for a while, but on 17th June 1971 a fire broke out just as the aircraft was taxying out for take-off; test pilot V. I. Mostovoy vacated the cockpit without even taking time to shut down the engines. The fire, which had been caused by a malfunction in the oxygen system, inflicted heavy damage on the aircraft which, though not a total loss, could only be repaired by the manufacturer in Novosibirsk.

Thus the State acceptance trials of the Su-15TM effectively stopped altogether; they resumed on 26th August when 'aircraft 0006' re-entered flight test after the Taifoon-M radar had been updated. In December it became clear that the rebuild of the first prototype was taking longer than expected and 'aircraft 0006' would have to shoulder the remainder of the State acceptance trials. The OKB and GK NII VVS attempted to widen the scope of the trials work by using the first two production Su-15TMs, '74 Blue' ('short' c/n 0302) and '75 Blue' (c/n 0303), which fully conformed to the 'Su-15 Stage II upgrade' standard as far as both airframe and avionics were concerned. The two aircraft arrived at Vladimirovka AB on 15th December; however, they had been handed over to the Air Force in such haste that the obligatory pre-delivery tests had not been performed and these had to be done at GK NII VVS instead of the factory. As a result, the two production Su-15TMs did not actually join the programme until March 1972. By then the rebuilt first prototype had returned to Vladimirovka AB, but it was a case of too little, too late – Stage A of the trials ended on 31st March.

The report on the results of Stage A said that the four participating aircraft had made 289 flights between them; of these, 81 flights (including 42 'for the record') were made under the Stage A programme proper and 91 (including 47 'for the record') under the 'preliminary report' test schedule. Nearly all performance targets had been met, with the exception of the weapons system's low-altitude performance. The maximum speed of the target being intercepted and the radar's target detection range were shorter than expected. The strong points noted by the pilots included the presence of the SAU-58 automatic flight control system and a short-range radio navigation system, both of which facilitated flying and landing, and, most importantly, the Su-15TM's greatly enhance combat potential as compared to the in-service Su-15 *sans suffixe*. The insufficient tailplane travel was cited as the main shortcoming. Eventually the strengths outweighed the weaknesses and the Su-15TM was recommended for production.

Stage B of the State acceptance trials began on 17th April 1972. The greater part of it was to be performed using the first two production Su-15TMs which were equipped with production Taifoon-M radars manufactured by LNPO Leninets, the other two aircraft being set aside for special test programmes. The radars proved fairly reliable; also, they featured a new module increasing the radar set's ECM resistance. At this stage the OKB had to tackle the complex task of increasing intercept efficiency at low altitudes (the aircraft did not meet the Air Force's requirements in this respect, and only one of five such missions performed by the end of April had been successful). A new GCI guidance algorithm had to be developed, and the result was felt immediately; 12 of the 16 low-altitude intercept missions that followed were successful, albeit they were flown over level terrain.

Stage B ended in April 1973, by which time a third production Su-15TM (c/n 0304) had joined the test fleet. All in all, the five aircraft made 256 flights, including 143 as part of the State acceptance trials programme (92 of them 'for the record') and another 103 under various special test programmes. Forty-six K-98M missiles were fired at La-17M, M-17 (MiG-17M), M-28 (IL-28M) and M-16 (Tu-16M) target drones, some of which featured active and passive ECM gear, as well as at RM-8 and PRM-1 paradroppable targets and KRM high-speed targets based on cruise missiles. Three of the target aircraft received direct hits and three others were destroyed by proximity detonation; in the other cases the missiles were instrumented inert rounds used for trajectory measurements.

The final report on the results of the State acceptance trials (Stage B) did not point out any major shortcomings; on the other hand, it effectively outlined a plan for the further upgrade of the Su-15TM. Among other things, in the future the aircraft was to be re-engined with the new Gavrilov R25-300 afterburning turbojets (see Su-15*bis* below). New wings and stabilators of increased area were to be incorporated, an automatic lateral stability augmentation system and a trim mechanism were to be provided.

The Su-15TM had entered production back in October 1971, superseding the Su-15T on the Novosibirsk production line; curiously, c/n 0302 was cited as the first production aircraft and no information exists on c/n 0301! Since the two models had a lot in common, the Su-15TM's batch numbering continued that of the Su-15T (ie, Su-15TM production started with batch 3). The in-house product code (*izdeliye* 37M or *izdeliye* 37 Srs M) remained unaltered. This was the last of the Su-15's production versions, the final 'TMs pertaining to batch 14 rolling off the production line in late 1975. Concurrently plant No.153 produced the Su-24 tactical bomber, hence the Su-15TM's production

Above: The first production Su-115TM ('74 Blue', c/n 0302) in its ultimate guise with an ogival radome and additional pylons for K-60 (R-60) 'dogfight missiles'; the stripes identify the K-60 as an inert round. The many access panels on the nose give it a patchwork appearance.

'75 Blue' (c/n 0303), the second production Su-115TM, during trials. The original conical radome has been replaced by the ogival version but the inner wing pylons have yet to be fitted. Note that the air intakes on top of the rear fuselage are still in place (compare with the aircraft in the upper photo).

rates were not particularly high, peaking at 110 aircraft per year.

The upgraded Su-15-98M aerial intercept weapons system was officially included into the inventory by a Council of Ministers directive dated 21st January 1975. It enabled manually- or automatically-controlled interception of single targets flying at altitudes of 500-24,000 m (1,640-78,740 ft) and speeds up to 1,600 km/h (990 mph) in pursuit mode and targets flying at altitudes of 2,000-21,000 m (6,560-68,900 ft) and speeds up to 2,500 km/h (1,550 mph) in head-on mode. The weapons control system had enhanced ECM resistance; on the down side, the service ceiling had decreased from 18,500 m (60,690 ft) – the figure obtained in the course of the State acceptance trials – to 17,970 m (58,960 ft). After service entry the Taifoon-M radar received the official designation RP-26, while the K-98M missile was renamed R-98M.

The first lot of production Su-15TMs was handed over to the military in the spring of 1972. Apart from the radar, production Su-15TMs were virtually identical to the Su-15T as regards avionics and equipment. Empty weight increased to 10,870 kg (23,960 lb) versus 10,220 kg (22,530 lb) for the Su-15 *sans suffixe*; the fuel capacity was 6,775 litres (1,490.5 Imp gal) and the fuel load 5,550 kg (12,235 lb).

One of the deficiencies the Sukhoi OKB and LNPO Leninets had to rectify together was the clutter on the radar display arising from internal reflections of the radar pulse inside the radome. The avionics house suggested fitting an ogival radome to cure the problem. After a series of tests on Su-15 c/n 1315340 at LII in June 1972 involving several differently shaped radomes such ogival radomes were fitted to the first two production Su-15TMs (c/ns 0302 and 0303) which underwent additional tests. The latter showed that the annoying echoes had vanished but the service ceiling, base altitude, effective range and interception range had decreased somewhat due to the extra drag created by the new radome. (Contrary to popular belief, the ogival radome was not a measure aimed at improving the Su-15TM's aerodynamics – in fact, it made them worse.)

Upon completion of the checkout trials at GK NII VVS in the autumn of 1973 the new radome was cleared for production. By then, however, a considerable number of Su-15TMs had been built with the old 'pencil nose'; new-build 'TMs received the ogival radome from batch 8 onwards and the previously manufactured aircraft were progressively updated in service.

The Su-15TM received the NATO reporting name *Flagon-F*.

The service tests of the new variant were held by one of the PVO's first-line units between 1st February 1975 and 20th July 1978 with good results. Like most aircraft, the aircraft was not immune to accidents; the first total loss was on 7th February 1973 when Su-15TM c/n 0820 suffered a critical failure during pre-delivery tests in Novosibirsk, forcing factory test pilot I. M. Gorlach to eject.

In the course of production the Su-15TM was constantly updated. Some of the major changes are listed below.

The engine nozzles' air ejector system identifiable by the characteristic air intake scoops on the aft fuselage was deleted from c/n 1015330 onwards and removed from previously manufactured aircraft in the course of overhauls. From batch 6 onwards the Su-15TM was equipped with the upgraded SAU-58-2 AFCS enabling automatic interception of low-flying targets (which was beyond the capabilities of the earlier SAU-58). The weapons complement was augmented with two R-60 'dogfight missiles' which, as already mentioned, had passed their trials in late 1974; the new weapon was integrated in situ from 1979 onwards.

From c/n 0911 onwards production Su-15TMs were equipped with new PU-2-8 launch rails which could be easily replaced with BD3-57M bomb racks (*bahlochnyy* derzhahtel' – beam-type rack), and appropriate changes were made to the electric system. Finally, at the insistence of the military the Taifoon-M radar was upgraded as part of the measures to offset the damage done by Lt. Viktor I. Belenko's notorious defection to Japan in a MiG-25P. The update was a success (see below) but its is not known with certainty if operational Su-15TMs were thus upgraded.

Su-15TM control system testbed (c/n 0303)

Service pilots kept complaining that that the Su-15's lateral stability was poor, especially during the landing approach. The Sukhoi OKB worked in several directions, trying to eliminate this shortcoming. Eventually the second production Su-15TM ('75 Blue', c/n 0303) was turned over to GK NII VVS for testing in 1974; this aircraft was retrofitted with a trim mechanism in the aileron control circuit and a lateral stability augmentation system. The new features received a positive appraisal and were introduced on the production line.

Later the same aircraft was used to test an increased-area horizontal tail. The test programme included 29 flights; the new horizontal tail was likewise recommended for production – too late, as Su-15TM production had ended by then; the change was not incorporated on operational aircraft.

Su-15TM strike armament testbed (c/n 1007)

The incorporation of bomb armament on the Su-15 for use as a strike aircraft had been repeatedly delayed ever since the beginning of the State acceptance trials. Finally, by August 1974 a production Su-15TM coded '76 Blue' (c/n 1007) was modified to feature four BD3-57M bomb racks instead of the standard pylons, with appropriate changes to the electrics. The K-10T sight was equipped with a tilting mechanism enabling attacks against ground targets.

The State acceptance trials proceeded in two stages; Stage A (October 1974 through May 1975) was to verify the bomb armament, while Stage B (June-December 1975) was concerned with the cannon and rocket armament. The aircraft was flown by GK NII VVS test pilots Ye. S. Kovalenko, V. A. Oleynikov and V. N. Moozyka. The trials were successful, the report stating that the armament fit under test rendered the aircraft suitable for use against pinpoint ground targets. As a result, late-production Su-15TMs were equipped with PU-2-8 launch rails which could be easily exchanged for BD3-57M racks.

Su-15TM avionics testbeds with modified radars (c/ns 1349 and 1407)

On 6th September 1976 Lt. Viktor I. Belenko, a pilot of the PVO's 530th IAP stationed in the Far East, defected to Japan in a MiG-25P, landing at Hakodate airport and laying the latest state-of-the-art in Soviet interceptor design wide open for inspection by the West. In response to this, the Soviet Council of Ministers issued a directive in November 1976 requiring measures to be taken in order to minimise the damage done by Belenko's treachery. The MiG-25P received the greatest attention, undergoing a complete change of the weapons control system which resulted in the advent of the MiG-25PD/MiG-25PDS. The Su-15TM, on the other hand, would make do with a modest upgrade of the radar.

In early 1977 two production Su-15TMs ('short' c/ns 1349 and 1407) were set aside for modification under this programme, undergoing tests in June-October that year. It took another 12 months to perfect the modified weapons control system, and in 1978 the modifications received approval for incorporation on in-service aircraft.

Su-15TM avionics testbed with Parol'-2 IFF

In keeping with the same 'anti-Belenko' directive of November 1976 another production Su-15TM coded '61' (c/n unknown) was taken from a service unit and modified for testing the new SRO-1P Parol'-2 (Password-2) IFF suite comprising an *izdeliye* 623-1 interrogator and an *izdeliye* 620-20P transponder. Outwardly the aircraft could be identified by a curved

Above: Su-15TM '76 Blue' (c/n 1007) was used to test the aircraft's suitability for the strike role. Here it is seen carrying 32-round UB-32A rocket pods on the wing pylons and UPK-23-250 cannon pods on the fuselage hardpoints.

Su-15TM '31 Blue' carries an improbable mix of ordnance, with a R-98MR on the port outer pylon, an R-98MT on the starboard outer pylon, two inert R-60s on the inner wing pylons and two 250-kg (550-lb) high-explosive bombs on the fuselage hardpoints.

This shot shows the same Su-15TM '31 Blue' armed with two R-98MTs and two inert R-60s taking off from a grass airstrip.

Table 1 shows the Su-15TM's interception ranges.

Interception range, km (miles)		Powerplant operation mode	
	cruise mode	full afterburner	full military power
head-on mode	504 (313)	360 (223)	300 (186)
pursuit mode	450 (279)	240 (149)	55 (34)

Tables 2 and 3 illustrate the specifications of the Su-15TM's missile armament.

	Launch range, km (miles)	Target elevation, m (ft)
R-98, head-on mode	8-18 (5-11)	3,000-4,000 (9,840-13,120)
R-98M, head-on mode	5-24 (3.1-14.9)	4,000-5,000 (13,120-16,400)
R-60, head-on mode	–	2,000 (6,560)
R-98, pursuit mode	2-14 (1.24-8.7)	–
R-98M, pursuit mode	2-15 (1.24-9.3)	–
R-60, pursuit mode	0.2-16 (0.12-9.93)	–
R-98, at low altitude	–	–
R-98M, at low altitude	2-3.5 (1.24-2.17)	–
R-60, at low altitude	0.3-1.5 (0.18-0.93)	–

	R-98	R-98M	R-60
Design g load	14	14	42
Operational on-wing g load	3	3	5

Table 4 provides a performance comparison of the radars fitted to the Su-15 and the Su-15TM.

	Su-15TM	Su-15
Radar type	Taifoon-M	Oryol-D58
Target detection range, km (miles):		
at high altitude	65 (40.3)	35 (21.7)
at low altitude	15 (9.3)	15 (9.3)
Target tracking range, km (miles):		
at high altitude	45 (27.9)	30 (18.6)
at low altitude	10 (6.2)	10 (6.2)
Scan limits:		
azimuth	±70°	±60°
elevation	+30°/–10°	+31°/–15°

antenna 'horn' positioned ventrally immediately aft of the radome and a probe aerial projecting aft from the dielectric fin cap. The modified aircraft underwent trials at GK NII VVS and the Parol' suite was recommended for production, becoming standard on Soviet military aircraft.

Su-15TM weapons testbed with internal cannon (c/n 0304)

The Sukhoi OKB did not give up on the idea of fitting an internal cannon to the Su-15; now a production GSh-23L cannon was to be mounted on the Su-15TM's fuselage underside aft of the nosewheel well à la MiG-23. The third production example (c/n 0315304), one of the five involved in the State acceptance trials, was modified for testing this cannon installation. The 'cannon saga' continued for almost three years, the modified aircraft eventually undergoing a separate State acceptance trials programme in 1973. The results were similar to those obtained with the standard UPK-23-250 pods; thus the built-in cannon was recommended for production. Yet the accuracy of the cannon fire was still rather poor because the standard K-10T sight was ill-suited for working with cannons, and, as a specialised gunsight could not be installed due to space limitations, the built-in cannon never found its way to the production line.

Su-15*bis* experimental interceptor (T-58*bis*)

On 25th February 1971, shortly after the completion of Stage A of the Su-15TM's State acceptance trials, the Council of Ministers issued a directive followed by a joint MAP/Air Force ruling. These documents required the Sukhoi OKB to re-engine the Su-15TM with Gavrilov R25-300 afterburning turbojets rated at 4,100 kgp (9,40 lbst) dry and 6,850 kgp (15,100 lbst) reheat, with a contingency rating of 7,100 kgp (15,650 lbst) up to an altitude of 4,000 m (13,120 ft).

Development work began in 1972, the aircraft receiving the in-house designation T-58*bis* and the provisional service designation Su-15*bis*. The prototype was converted from the fifth production Su-15TM coded '25 Blue' (c/n 0306) at the Novosibirsk aircraft factory in the first half of the year, making its maiden flight on 3rd July with Sukhoi OKB chief test pilot Vladimir S. Ilyushin at the controls. V. Vasil'yev was engineer in charge of the tests.

The manufacturer's flight tests continued until 20th December; the aircraft was also flown by OKB test pilots A. N. Isakov and V. A. Krechetov. As compared to the standard R13-300 powered aircraft the Su-15bis had markedly better acceleration characteristics and a higher top speed at low and medium altitudes, especially with the engines at

This Su-15TM served as a testbed for the Parol'-2 IFF transponder. The non-standard IFF aerials project forward from beneath the radome and aft from the top of the fin. The production version of the system featured triangular blade aerials. Note the 'Excellent aircraft' maintenance award badge rarely seen on Su-15s.

contingency rating. Service ceiling and interception range in head-on mode were also improved a little.

The Su-15bis underwent State acceptance trials between 5th June and 10th October 1973; GK NII VVS test pilots N. I. Stogov, V. I. Mostovoy and Ye. S. Kovalenko made a total of 79 flights. The trials report said that performance below 4,000 m was improved over the standard Su-15TM and using the contingency rating (or rather power boost perhaps) allowed targets flying at up to 1,000 km/h (620 mph) to be intercepted in pursuit mode. The Su-15bis was recommended for production but, ironically, it never achieved production status. Conversely, the R25-300 engine did, powering the mass-produced MiG-21bis tactical fighter.

Su-15UM combat trainer (U-58TM, *izdeliye* 43)

In late 1974 the OKB started design work on a new combat trainer which received the manufacturer's designation U-58TM. Developed to meet an Air Force requirement for a trainer version of the Su-15TM, the aircraft was intended for training PVO pilots in flying techniques, aerobatics and combat tactics. In accordance with a joint MAP/MRP ruling agreed on by the Air Force the Sukhoi OKB was to develop a set of project documents for a combat trainer version of the Su-15TM within the shortest possible time, whereupon the Novosibirsk aircraft factory would build a prototype. The project documents were transferred to plant No.153 in the spring of 1975, but it was not until 16th July 1975 that MAP finally issued an order officially initiating development of the U-58TM.

The parties concerned chose not to build a prototype, performing the required trials on the first production aircraft. The trainer was based on the airframe of the late-production Su-15TM; unlike the Su-15UT, there was no fuselage stretch, the overall length being the same as the single-seater's. The twin-wheel nose gear unit was also retained.

Remarkably, the provision of a second cockpit did not incur a reduction of the fuel tankage, the space for it being provided solely by deleting some equipment items. The internal fuel capacity was 6,775 litres (1,490.5 Imp gal) and the fuel load 5,550 kg (12,235 lb); additionally, two 600-litre (132 Imp gal) drop tanks could be carried.

In addition to a full set of controls and flight instruments, the rear cockpit featured a special control panel allowing the instructor to deactivate some of the instruments in the trainee's cockpit, simulating hardware failures. The canopy was similar to that of the Su-15UT, the rear section incorporating a retractable forward vision periscope. The area of the horizontal tail was increased slightly to address the elevator authority problem typical of the Su-15.

Equally remarkably, the U-58TM's empty weight of 10,635 kg (23,445 lb) was lower than the single-seater's. This was accom-

Table 5 provides a performance comparison of Soviet interceptors of the 1970s.

	Su-15TM	Su-15	MiG-25PD/MiG-25PDS
Indicated airspeed at sea level, km/h (mph)	1,300 (807)	1,100 (683)	1,200 (745)
Maximum true airspeed at 11,000 m (36,090 ft), km/h (mph)	2,230 (1,385)	2,230 (1,385)	3,000 (1,863)
Service ceiling, m (ft)	18,000 (59,050)	17,650 (57,900)	20,500 (67,260)
Endurance	1 hr 41 min	1 hr 27 min	1 hr 37 min
Base altitude:			
head-on mode	17,000 (55,770)	15,000 (49,210)	20,000 (65,620)
pursuit mode	16,000 (52,500)	n.a.	n.a.
Climb time to base altitude, minutes	n.a.	n.a.	9.7
Maximum range, km (miles)	1,380 (860)	1,120 (695)	1,320 (820)
Operational g limit	+6.5	+6.0	+4.4
Maximum thrust/weight ratio	0.82	0.63	0.7
Wing loading, kg/m² (lb/sq.ft)	397 (81.39)	395 (80.98)	490 (100.46)

Table 6 compares the performance of some Soviet and US interceptors.

	MiG-25P	MiG-23M	Su-15TM	McDD F-15A	GD F-16A
Thrust/weight ratio	0.66	0.85	0.7	1.15	1.4
Top speed, km/h (mph):					
at sea level	1,200 (745)	1,350 (838)	1,300 (807)	1,470 (913)	1,490 (925)
at 11,000 m	3,000 (1,863)	2,500 (1,550)	2,230 (1,385)	2,650 (1,645)	2,100 (1,300)
Maximum Mach number	2.83	2.35	2.1	2.5	2.0
Service ceiling, m (ft)	20,500 (67,260)	17,800 (58,400)	18,000 (59,050)	19,200 (62,990)	18,000 (59,050)
Minimum control speed, km/h (mph)	400 (248)	400/450 (248/279)	450 (279)	350 (217)	350 (217)

plished by deleting much of the Su-15TM's avionics, namely the radar, the SAU-58-2 AFCS, the Lazoor'-M command link system, the SPO-10 radar warning receiver and the RSBN-5S SHORAN. The avionics included an R-832M radio, an ARK-10 ADF, an RV-5 low-range radio altimeter, an MRP-56P marker beacon receiver, a KSI-5 compass system and an AGD-1 artificial horizon, as well as an SPU-9 intercom and an MS-61 CVR.

The customer required the trainer to retain a measure of combat capability so that weapons training could be performed. Given the lack of radar, this requirement could be met by using heat-seeking missiles; thus the U-58TM featured a weapons control system compatible with the R-98T medium-range AAM and the R-60 short-range AAM. Additionally, UPK-23-250 pods could be carried on the fuselage hardpoints.

The 'second-generation' trainer variant received the service designation Su-15UM (*oochebnyy, modernizeerovannyy* – trainer [version], upgraded). Aptly coded '01', the production prototype of the Su-15UM ('short' c/n 0101) was rolled out at Novosibirsk-Yel'tsovka in the spring of 1976; Yuriy K. Kalintsev was appointed engineer in charge of the tests. The maiden flight took place on 23rd April with factory test pilots V. T. Vylomov and V. A. Belyanin at the controls. Five days later the aircraft was flown to Zhukovskiy where OKB test pilots Yevgeniy S. Solov'yov and Yuriy A. Yegorov performed an abbreviated manufacturer's test programme comprising only 13 flights.

On 23rd June the Su-15UM production prototype was turned over to the military for

Left: Head-on view of a Su-15TM in definitive production configuration with ogival radome and four missile pylons.

Below: A typical late-production Su-15TM. Apart from the radome, late-production examples were characterised by the absence of the large air intake scoops on top of the rear fuselage.

Bottom: This view illustrates the Su-15TM's wing planform. Note the anti-glare paint on the air intakes.

State acceptance trials; these were rather brief, lasting only five months, and were completed on 25th November with good results. At this stage the aircraft was flown by GK NII VVS test pilots Ye. S. Kovalenko, V. V. Kartavenko, Oleg G. Tsoi and V. Ye. Kostyuchenko who made a total of 72 flights, including 60 'for the record'. The State commission's report said that the new two-seater was suitable for training aircrews in take-off and landing techniques, all flight elements and certain elements of combat tactics; the Su-15UM was recommended for production and service.

The Novosibirsk aircraft factory produced the new trainer in 1976-1980 under the product code *izdeliye* 43 (indicating it was a follow-on to the Su-15UT which was *izdeliye* 42). The type was cleared for service without the usual Council of Ministers directive – a Ministry of Defence order was all it took.

The Su-15UM's NATO reporting name was *Flagon-G*.

The only major change made to the Su-15UM in the course of production was that the RSBN-5S SHORAN was added after all; new-build examples were equipped with the system from batch 4 onwards and deliveries commenced in 1978. The final three trainers off the line were delivered fairly late; two of them ('short' c/ns 0443 and 0444) left the plant in February 1981, while Su-15UM c/n 0445 – the very last *Flagon* built – did not make its first flight until 14th February 1982, a full year after the rollout. The honour of making the 'last first flight' of a Su-15 fell to factory test pilots I. Ya. Sushko and Yu. N. Kharchenko.

T58D-30 interceptor (project)

In February 1966 the Sukhoi OKB considered re-engining the Su-15 with D-30 turbofans. Developed by Pavel Aleksandrovich Solov'yov's OKB-19 in Perm' for the Tu-134 short/medium-haul airliner, the commercial version of this engine was rated at 6,800 kgp (14,990 lb st), although an even more powerful afterburning version was probably envisaged for the fighter. However, installing D-30s would require major structural changes,

Above: The sole Su-15*bis* (c/n 0306) seen during trials. The tactical code '25 Blue' was very probably a hint at the engine type (R25-300). Note the Sukhoi OKB's 'winged archer' logo beneath the the cockpit.

Another view of the Su-15*bis*. The aircraft was based on an early-production airframe and thus featured the nozzle cooling air scoops on top of the rear fuselage.

which would disrupt production, and the project – provisionally designated T58D-30 to indicate the engine type – was shelved.

T-58Sh attack aircraft (project)

In the summer of 1969 the Sukhoi OKB looked into the possibility of transforming the T-58 into a fully-fledged attack aircraft. The idea was triggered by engineer Oleg S. Samoylovich's initiative to design an all-new ground attack aircraft which was initially known as the SPB (*samolyot polya boya* – battlefield aircraft) before becoming the T-8 – ie, the famous Su-25. The SPB project unexpectedly encountered competitor within its own OKB: A. M. Polyakov, the chief project engineer of the Su-7 and Su-17, proposed an attack version of the T-58 interceptor designated T-58Sh (*shtoormovik* – attack aircraft).

Billed as an 'in-depth upgrade' of the interceptor, the T-58Sh was for all intents and purposes a new aircraft. The forward fuselage up to frame 10 was new, being drooped to provide adequate downward visibility. The wings were also new, featuring a trapezoidal planform with reduced leading-edge sweep and greater area; the overall appearance was not dissimilar to the Dassault Étendard. The cockpit section and the engines were protected by armour and the fuel tanks were self-sealing for higher survivability. The radar and other mission avionics associated with the

'01 Blue', the production prototype of the Su-15UM trainer. The proportions of the Su-15UM make an interesting comparison with those of the Su-15UT (see page 69) – note the position of the forward cockpit relative to the intakes.

Above: This diagram of the projected Su-19M illustrates the ogival wings, the six wing hardpoints, the two fuselage hardpoints and the built-in cannon on the centreline.

interceptor role were replaced by an ASP-PF gunsight, a PBK-2 bomb sight optimised for lob-bombing (pri-**tsel** dlya bombome**ta**niya s kabree*rovaniya*) and a Fon (Background) laser rangefinder.

The aircraft featured eight weapons hardpoints and a built-in *izdeliye* 225P Gatling cannon. The external stores options included bombs of up to 500 kg (550 lb) calibre, unguided rockets of assorted calibres, Kh-23 air-to-surface missiles, UPK-23-250 cannon pods and SPPU-17 pods with movable cannons for strafing ground targets in level flight. For self-defence the T-58Sh would be armed with K-55 and K-60 AAMs. At a 17.5-ton (38,580-lb) take-off weight the aircraft was to have a 4-ton (8,820-lb) ordnance load.

Eventually, however, the T-8 was selected as the more promising design; the T-58Sh did not progress beyond the general arrangement drawing stage.

Su-19 (T-58PS) advanced interceptor (project)

In 1972-73 the Sukhoi OKB proposed an in-depth upgrade of the Su-15, striving to enhance the interceptor's performance by radically improving the aerodynamics. Since the OKB placed high hopes on the ogival wings developed for the original T-10 *Flanker-A* fighter (the precursor of the Su-27), the intention was to use such wings on the Su-15 (T-58) as well. The rewinged interceptor bore the in-house designation T-58PS; the meaning of the suffix is unknown.

A series of wind tunnel tests was held at TsAGI, followed by more detailed research into layouts utilising ogival wings. Estimates showed that the interceptor's performance and agility would be enhanced dramatically; also, the new wings provided room for two additional hardpoints, allowing more short-range AAMs to be carried – a real asset in a dogfight, in which the more manoeuvrable fighter could now engage.

Later the T-58PS was referred to in the Sukhoi OKB's official correspondence with MAP and the Air Force as the Su-19. Interestingly, Western intelligence agencies somehow got wind of this designation and erroneously attributed it to the Su-24

Su-19M advanced interceptor (project)

The next step towards improving the performance of the prospective Su-19 (T-58PS) interceptor was to install advanced R67-300 engines instead of R13-300s. The Sukhoi OKB prepared a technical proposal for an R67-300 powered version, sending it to MAP and the Air Force for appraisal. According to this document the rewinged aircraft (ie, the basic Su-19) could enter flight test in the fourth quarter of 1973, the re-engined version designated Su-19M following in the first quarter of 1975. Yet the military showed a complete lack of interest; whet they needed more was a new radar giving the fighter 'look-down/shoot-down' capability.

Su-15M advanced interceptor (project)

Failing to get the go-ahead for Su-19/Su-19M development, the OKB then proposed fitting the existing Su-15TM with a new *Poorga* (Snowstorm) fire control radar to give it the required 'look-down/shoot-down' capability. The PVO top command, which favoured the Su-15, supported this idea. In addition to the new radar, the proposed upgrade (provisionally designated Su-15M) involved installation of Lyul'ka AL-21F-3 afterburning turbojets rated at 7,800 kgp (17,195 lbst) dry and 11,215 kgp (24,725 lbst) reheat and integrating K-50 AAMs, four of which were to be carried. (According to other sources the Su-15M was to be armed with K-25 AAMs.) Yet MAP regarded this project with a jaundiced eye and all further work on upgrading the Su-15 had to be abandoned.

Two views of the proposed T-58Sh attack aircraft which lost out to the T-8. It is easy to see that the aircraft has little in common with the Su-15 – indeed, it looks more like the Dassault Etendard than anything else!

Chapter 6

'Target in Sight!'

Su-15 in Action

On the morning of 9th July 1967 huge crowds of Muscovites and visitors to the capital rushed to Moscow's Domodedovo airport where a spectacular airshow – the first in the exactly six years – was to take place. After the hardships of the Khrushchov era the Soviet aircraft industry and the Air Force were eager to show the nation's new leaders and the public at large that, in spite of the battering they had taken due to Khrushchov's 'missile itch', they were still very much alive and had considerable potential. Hence it was decided to display almost everything the industry had to offer, including aircraft that were still undergoing trials at the time or were just about to enter production and/or service. The latter included the Sukhoi OKB's latest product, the Su-15 interceptor, which received the honour of opening the flying display of the show. Right on schedule a group of five Su-15s flown by 148th TsBP i PLS pilots and led by Col. P. P. Fedoseyev made a high-speed pass over the improvised grandstand and pulled into a spectacular formation climb, fanning out at the top. The display also included a short take-off and landing demonstration by the experimental T-58VD piloted by OKB test pilot Yevgeniy S. Solov'yov. A while later, Sukhoi OKB chief test pilot Vladimir S. Ilyushin made a flypast in a production Su-15 coded '47 Red' and painted black overall for sheer effect.

From the number of Su-15s participating in the show Western military analysts concluded that the Soviet Union had fielded a new interceptor, and the Su-15 was allocated the NATO reporting name *Flagon*. Western aviation experts made a fairly accurate guess as to the fighter's performance and correctly guessed that the aircraft was powered by Tumanskiy R11 afterburning turbojets. The advent of the Su-15TM with its ogival nose, however, confused the Western experts somewhat; for some reason they decided that the aircraft was fitted with a new radar (which was correct) and powered by Lyul'ka AL-21F engines (which was absolutely wrong). Codenamed *Flagon-F*, the upgraded aircraft was referred to in the Western press as the 'Su-21'; it was quite some time before the correct designation became known.

As already mentioned in the previous chapter, the Su-15-98 aerial intercept weapons system was formally included into the Soviet Air Defence Force inventory in April 1965 upon completion of the State acceptance trials. The system was capable of intercepting targets flying at speeds of 500-3,000 km/h (310-1,860 mph) and altitudes of 500-23,000 m (1,640-75,460 ft). The interceptor was guided towards the target by the Vozdookh-1 automated GCI system until the target came within range of its fire control radar.

The top brass of the Soviet Ministry of Defence and the PVO had a lot riding on the new interceptor which was to replace several obsolete aircraft types in the PVO inventory. Following the usual practice, the 148th TsBP i PLS at Savostleyka AB was the first unit to master the new type; the Centre's 594th UIAP (*oochebnyy istrebitel'nyy aviapolk* – fighter training regiment) started conversion training for the Su-15 in early 1966. Su-15 production in Novosibirsk was getting under way slowly, and the practice part of the training course could not begin until Sukhoi OKB test pilot Vladimir S. Ilyushin had ferried the second pre-production aircraft (c/n 0015302) from Zhukovskiy to Savostleyka AB on 28th October 1966. By the end of the year the Centre's pilots had made 14 flights in this aircraft.

Deliveries of truly production Su-15s to the 594th UIAP began in January 1967. Since no dual-control trainer version of the Su-15 existed yet, conversion training had to be undertaken using single-engined Su-9U trainers. Even so, the Su-9U was in short supply and was badly needed by the operational units flying the type, and the decision was taken to use swept-wing Su-7U trainers instead. In May 1967 service pilots started taking their conversion training in the Su-7U before making their first flights in the Su-15.

In the spring of 1967 a display team was formed at the 148th TsBP i PLS specially for

Above: This Su-15 painted black overall to create a menacing effect and coded '47 Red' gave a solo performance at the Moscow-Domodedovo air show on 9th July 1967.

Three of the five Su-15s that opened the flying display at the 1967 show. The fighters carry dummy R-98 missiles.

Above: A Su-15 coded '01 and carrying two UPK-23-250 cannon pods streams its brake parachute after landing.

Above: A still from a Soviet documentary showing a Su-15TM firing a radar-homing R-98MR missile.

'No malfunctions.' A Su-15 pilot writes his comments after a sortie for the aircraft's crew chief. The legend on the nose reads 'Otlichnyy' (Excellent), a maintenance award; it was later replaced by a badge.

the abovementioned airshow at Moscow-Domodedovo which was to take place in July; the team included 594th UIAP pilots, pilots from the Centre's Command & Control Squadron and a few service pilots. To rehearse the display the team temporarily relocated to Zhukovskiy together with its aircraft. Training flights were made with dummy R-98 missiles and did not go without incident; on 4th July, five days before the show, one of the fighters (c/n 0315304) lost a missile together with the pylon after pulling into a step climb. Investigation of the incident showed that the aircraft had exceeded its operational G load limit by far, and it was no wonder that the pylon had broken off. Detailed examination of the airframe revealed substantial permanent structural deformation and the aircraft was declared a write-off; thus the whole affair was actually a non-fatal accident. To prevent further incidents the actual display flight at the show was performed without missiles and with smoke generator pods on the fuselage hardpoints instead of drop tanks.

Deliveries to operational PVO units began in the spring of 1967. The 611th IAP of the Moscow PVO District based at Dorokhovo AB (Yaroslavl' Region) was the first to re-equip. The 62nd IAP based at Bel'bek AB on the Crimea Peninsula, the Ukraine, followed in July, the 54th GvIAP at Vainode AB, Estonia, re-equipping shortly afterwards.

Under the conversion training programme one pilot for each Su-15 would be trained within a six-month period to perform combat duty in visual and instrument meteorological conditions in the daytime and in VMC at night. This proved difficult to accomplish; to hasten the training process, qualified flying instructors were seconded to operational PVO units from the 148th TsBP i PLS.

The strategic bombers in service with the US Air Force (primarily the Boeing B-52 Stratofortress) and the Royal Air Force (the V-Bombers), as well as the Hound Dog (USAF) and Blue Steel (RAF) supersonic air-to-surface missiles carried by these aircraft, were envisaged as the principal targets which the Su-15 would have to deal with. As a dogfighting machine the Su-15 was no good, of course, as it lacked the agility – but then, it was not designed with dogfights in mind. The addition of R-60 heat-seeking short-range AAMs did not increase the interceptor's chances in the event of an encounter with enemy fighters but still improved the chances of a 'kill' against a typical target.

By June 1968, 130 Su-15s were in service with eight fighter regiments. A total of 149 pilots were qualified to fly the type but less than 50% of them were fully trained to perform combat duty in IMC in the daytime and only two (!) were able to fly night sorties. During training special attention was given to engagements in head-on mode, as this type of attack was new for Soviet interceptor pilots. Live weapons training at a weapons range near Krasnovodsk involving missile launches commenced in April 1967; the 611th IAP's pilots were the first to do so, firing 47 missiles at various practice targets.

The same 611th IAP was selected to hold the obligatory service trials (evaluation) of the new interceptor, receiving ten Su-15s (mostly Batch 3 aircraft). The trials proceeded from 29th September 1967 to 15th May 1969. During this period the ten fighters made a total of 1,822 flights, including 418 under the actual evaluation programme; two live weapons training sessions were held with the expenditure of 58 AAMs. The service tests basically corroborated the results of the State acceptance trials. However, a number of serious shortcomings was discovered; among other things, the service ceiling fell short of the specifications due to a decrease in engine thrust in the course of the engines' service life, and the interception range was also less than expected.

By the end of 1975 the PVO intended to re-equip 41 fighter regiments with the Su-15, whereupon the new interceptor would make up nearly 50% of the PVO's aircraft fleet. However, when Su-15 production ended in early 1976, the type was in service with only 29 units, 18 of them operating the Su-15 *sans suffixe* and a further eleven units flying the Su-15TM.

To give credit where credit is due, of all fighter types operated by the PVO the Su-15 probably had the highest percentage of successful real-life intercepts of aircraft intruding into Soviet airspace. Its baptism of fire came on 11th September 1970… well, actually the expression 'baptism of fire' is not really applicable because no shots were fired on this occasion. At 03:36 Moscow time PVO radar pickets near Sevastopol', the Ukraine, detected a lone aircraft heading north towards the Soviet border at 3,000 m (9,840 ft), and a 'Red Alert' was called. The target was then 260 km (160 miles) south-west of the city; when it approached within 100 km (62 miles) of the border, a 62nd IAP Su-15 scrambled from Bel'bek AB to prevent an incursion. The target turned out to be an elderly Douglas C-47 belonging to the Greek Air Force, and when it eventually crossed the border the fighter lined up alongside and rocked its wings in the internationally recognised 'follow me' signal. The Dakota complied, landing at Bel'bek AB. It turned out that the pilot, Lt. M. Maniatakis, had stolen the aircraft from Kania AB on the island of Crete and fled from his homeland where the fascist junta of the Black Colonels had seized power. Maniatakis requested political asylum in the USSR, which was in all probability granted.

Above: A Su-15 coded '01 Blue' and carrying zebra-striped dummy R-98s is refuelled on the taxiway of a snowbound Soviet airbase.

Throughout the 1970s the southern borders of the Soviet Union perpetually received the attentions of hostile aircraft coming from Turkey and Iran. The events described below are but a few of the incursions that took place there.

On 7th September 1972 a flight of Turkish Air Force (THK – Türk Hava Kuvvetleri) North American F-100 Super Sabres entered Soviet airspace near Leninakan, Armenia (the city is now called Gyumri). Despite flying at ultra-low altitude, the intruders were detected by air defence radars in a timely fashion. Another ploy of the 'bad guys' worked, however – the fighters flew in close formation, appearing on the radarscopes as one heavy aircraft (the USAF had used this tactic in Vietnam); hence only a single 166th IAP Su-15 was scrambled from Sandar AB in neighbouring Georgia to intercept 'it'. The GCI command post operators did not realise that the target was not an 'it' but a 'they' until the Turkish fighters swept over the place with a roar.

The lone Su-15 proved incapable of intercepting its quarry because its radar lacked 'look-down/shoot-down' capability. As a result, the F-100s passed over Leninakan and were fired upon by a heavy machine gun providing anti-aircraft protection for the PVO's radar site but got away unscathed.

On 23rd May 1974 another THK F-100 intruded into Soviet airspace over the Caucasus region with impunity. A Su-15 standing on QRA duty scrambled from the airbase in Kyurdamir, Azerbaijan, but was not directed towards the target because the latter had unwisely intruded in an area defended by an SAM regiment. A missile was fired at the F-100 but missed due to a malfunction in the guidance system.

Eventually, however, the Turks fell victim to the rule 'pride goeth before the fall'. On 24th August 1976 Soviet AD radars detected a target moving in Turkish airspace towards the Soviet border. This was soon identified as a pair of F-100s flying in close formation.

A Polish officer examines the cockpits of a visiting Su-15UT as his colleagues watch a Su-15TM coming in to land during one of the type's few foreign deployments.

No fewer than three Su-15s scrambled this time (two from Kyurdamir and one from Sandar AB), but again they did not manage to get a piece of the action. The fighters had again rashly flown right into a nest of SAMs; this time the PVO crews on the ground did their job well and one of the Super Sabres was shot down. Unfortunately the wreckage fell on the wrong side of the border and the pilot, who ejected, also landed in Turkish territory; the following day the Turks raised hell, accusing the Soviet Union of the 'wanton destruction of a Turkish fighter'.

A while earlier, on 2nd April 1976, a 777th IAP Su-15 flown by Lt (sg) P. S. Strizhak scrambled from Sokol AB on Sakhalin Island to intercept a USAF Boeing RC-135 reconnaissance aircraft which had entered the 100-km territorial waters strip. Shortly after take-off the pilot was redirected towards a new target – a Japanese Maritime Self-Defence Force (JMSDF) Lockheed-Kawasaki P2V Neptune reconnaissance aircraft flying over the Sea of Japan at 2,000 m (6,560 ft) off the southern tip of Sakhalin, a militarily sensitive area. Approaching within 5-6 km (3.1-3.7 miles) of the target, the interceptor followed it, flying a parallel course. Apparently Strizhak flipped the wrong switch and inadvertently fired an R-98R missile at the Neptune, though no order to attack had been given. Realising what he had done, the pilot made a turn just in time, causing the missile to lose target lock-on; the missile passed off the spyplane's starboard wing and self-destructed.

At 14:57 Moscow time on 23rd December 1979 a Cessna 185 Skywagon light aircraft entered Soviet airspace 175 km (108 miles) south-west of Maryy, Turkmenia (pronounced like the French name Marie), coming from Iranian territory and flying at about 3,000 m. The aircraft was detected by PVO radars three minutes before it crossed the border, and a 156th IAP Su-15 took off almost immediately from Maryy-2 AB to intercept it. The pilot was directed towards the target by GCI stations but failed to spot it because the Cessna was camouflaged (so much for allegations about 'navigation errors'). The radars lost track of the target shortly afterwards and, after circling for a few minutes, the fighter pilot had no choice but to head for home. (Three more Su-15s and a MiG-23M had also scrambled by then, but they were not directed towards the target.) Nevertheless, his mission was accomplished; when (unbeknownst to the Soviet pilot) the interceptor passed directly above the Cessna, its pilots aborted their plan, losing altitude and opting for an emergency landing for fear of being shot down (hence the disappearance of the target from the radarscopes). Eventually they landed on a highway 195 km (121 miles) west of Maryy and were soon arrested by Soviet border troops.

Another incident on the Iranian border occurred on 18th July 1981. An unidentified aircraft flying at about 8,000 m (26,250 ft) briefly entered Soviet airspace but then left it and the pair of 166th IAP Su-15s which had taken off to intercept it was ordered back to base. A few hours later, however, another unidentified aircraft intruded into Soviet airspace; this time a single 166th IAP Su-15 flown by Capt. V. A. Kulyapin and armed with two R-98s and two R-60s scrambled to intercept it. The intruder turned out to be a Canadair CL-44 freighter leased from an Argentinean airline and flown by a Swiss crew. The fighter pilot gave the customary 'follow me' signals, trying to force it down at a Soviet airfield; instead, the big turboprop started manoeuvring dangerously, making sharp turns in the direction of the Su-15. The pursuit continued for more than ten minutes; eventually Kulyapin received orders to destroy the intruder. Since the border was very close and the target could get away before the fighter could move away to a safe distance for missile launch, Kulyapin chose to ram the target. The attack was skilfully executed; moving into line astern formation, the Su-15 pitched up into a climb, slicing off the CL-44's starboard tailplane with its fin and fuselage. The freighter plummeted to the ground, killing all on board; however, Kulyapin's aircraft was seriously damaged by the collision and the pilot ejected, landing safely not far from the wreckage of both aircraft. This time the wreckage fell on Soviet territory, furnishing irrefutable evidence of the border violation. For this act Capt. Kulyapin was awarded the Order of the Red Banner.

Gradually, together with the MiG-25P heavy interceptor capable of Mach 2 flight, the Su-15 supplanted the outdated Su-9, Su-11, Yak-28P and MiG-21PFM from the PVO inventory. Su-15s saw service with units stationed in almost all borderside regions of the Soviet Union, the High North and the Far East receiving the highest priority. The Su-15TM which superseded the initial versions on the production line remained one of the principal fighter types defending these vital areas for many years. The upgraded Su-15-98M aerial intercept weapons system comprising this aircraft and the Vozdookh-1M GCI system which permitted guidance in manual, semi-automatic (flight director) and fully automatic modes was capable of intercepting targets flying at speeds of 500-2,500 km/h (310-1,550 mph) and altitudes of 500-24,000 m (1,640-78,740 ft).

'10 Blue', a late-production Su-15TM in service with one of the PVO's first-line units.

The Su-15TM also saw a good deal of action in defence of the Soviet borders, particularly in the late 1970s and the 1980s. 20th April 1978 was the first occasion when a South Korean aircraft 'accidentally' strayed into Soviet airspace. The full truth about this incident remains unknown to this day. Some Western media maintain that the incursion was a result of crew error because the pilots were making their first flight in an unfamiliar aircraft along an unfamiliar route. Get real. It is hard to imagine a navigation error that would lead to a course change in excess of 100°.

The facts: at 20:54 Moscow time the radar pickets of the 10th Independent PVO Army detected an aircraft 380 km (236 miles) north of Rybachiy Peninsula, flying at 10,000 m (32,800 ft) and heading towards Soviet territorial waters at about 900 km/h (559 mph). When the target approached the 100-km territorial waters strip, at 21:11 the officer of the day at the 10th Independent PVO Army headquarters ordered a scramble. Since the unit based nearest to the coast was re-equipping with new aircraft and was not operational for the time being, the mission fell to the 431st IAP at Afrikanda AB (Arkhangel'sk Region), and a Su-15TM piloted by Capt. A. I. Bosov took off to intercept the target. After being directed towards the unknown aircraft in head-on mode by GCI control the pilot reported seeing the target on his radar display, executed a port turn and started closing in on the target. Coming within visual identification range, Bosov reported it was a four-engined Boeing 747 (sic) but he could not make out the insignia – they were either Japanese, Chinese or Korean. (Obviously the pilot had seen hieroglyphic characters painted on the aircraft's fuselage but had no way of knowing what language it was – Auth.)

Above: This Su-15TM coded '09 Blue' was operated by one of the PVO's schools. Note the missile launch rails on the inboard pylons.

Actually this was no 747 but a Korean Air Lines Boeing 707-321BA-H registered HL7429 (c/n 19363, fuselage number 623) bound from Paris-Orly to Seoul on flight KE902. As David Gero wrote in his book Flights of Terror – Aerial Hijack and Sabotage since 1930, 'Built more than a decade earlier, the aircraft lacked a modern inertial navigation system, and as a magnetic compass is useless in this part of the world (it gives false readings due to the proximity of the North Pole – Auth.), and with a scarcity in ground aids, the crew would have to rely upon the older but well-proven method of celestial navigation.

Trouble first arose in the vicinity of Iceland, when atmospheric conditions prevented the aircraft from communicating with the corresponding ground station. Approximately over Greenland, and following the instructions of the navigator, the 707 **inexplicably initiated a turn of 112 degrees,** *heading in a south-easterly direction towards the USSR* (my highlighting – Auth.). *A while later the pilot, Captain Kim Chang Kyu, sensed something was amiss by the rather obvious fact that the sun was on the wrong side of the aircraft!'*

Capt. Bosov was instructed to force the intruder down at a Soviet airfield, which he tried to do, making two passes along the 707's port side 50-60 m (165-200 ft) away to a point ahead of the flightdeck and rocking the wings. Yet the South Korean crew ignored these 'amorous advances'.

Meanwhile, after analysing the target's track plotted by AD radars, the 10th Independent PVO Army HQ decided the 707 was pressing on towards the Finnish border, which was only five minutes away, in an attempt to escape and ordered the airliner shot down. At 21:42 the fighter pilot fired a single R-98MR missile, reporting an explosion and saying that the target was losing altitude; Bosov was about to fire a second missile but lost target lock-on because the Boeing was descending sharply.

In the meantime a steady exchange of information was going on between PVO command centres at all echelons. The PVO

Two Su-15UMs await the next training session at a hardstand equipped with a centralised electric power distribution system obviating the need for power carts. Note the different tactical code presentation on aircraft obviously belonging to the same unit.

85

Commander-in-Chief was belatedly informed that the target was a civil airliner; hence the C-in-C's order not to shoot the intruder down but to force it down in one piece reached the lower echelons too late, when the 707 was already under attack. The explosion tore away the Boeing's port wingtip and aileron, knocked out the No.1 engine and apparently punctured the fuselage, causing a decompression. The crew initiated an emergency descent, causing the PVO radar pickets to briefly lose sight of the aircraft. By then, apart from Bosov's Su-15, five other aircraft had scrambled to intercept the intruder – two Yak-28Ps from Monchegorsk, one MiG-25P from Letneozyorsk, a Su-15TM from Poduzhem'ye AB and a further Su-15TM from Afrikanda AB. When the target vanished from the radarscopes, a further Yak-28P, a MiG-25P and three Su-15TMs from the same bases joined the hunt. A 265th IAP Su-15TM from Poduzhem'ye even fired a missile at a slow-flying target at 5,000 m (16,400 ft) – which later turned out to be nothing more than a cloud of honeycomb filler fragments from the 707's damaged wing.

The crippled airliner circled at low altitude near Loukhi settlement near Kem', Arkhangel'sk Region, where it was again detected and tracked by AD radars and the nearest interceptor was directed to the scene. Since the Su-15's radar was not much use against a low-flying target, the pilots had to rely in the Mk 1 eyeball; yet mortal men haven't got the eyesight of an owl, and even on a cloudless polar night it takes time to locate the target. At 22:45 Capt. Keferov of the 265th IAP spotted the intruder flying at 800 m (2,620 ft) near Loukhi; 12 minutes later the target was spotted by another 265th IAP pilot, Maj. A. A. Ghenberg. Together they gave signals to the crew, trying to force the jet to follow them; the airliner ignored the signals, landing on the frozen Lake Korpijärvi 5 km (3.1 miles) south-west of Loukhi. Of the 109 occupants, two passengers were killed (allegedly by fragments of the damaged engine) and 13 people were injured. The crew and passengers of the 707 were detained by the Soviet authorities but subsequently released; the airliner, which was declared a write-off, was recovered from the scene and taken to Moscow for examination.

A much more tragic incident with far-reaching political consequences took place on the night of 1st September 1983. Another Korean Air Lines aircraft, Boeing 747-230B HL7442 (c/n 20559, f/n 186), entered Soviet airspace en route from Anchorage, Alaska, to Seoul on flight KE007 (an apt flight number, isn't it?). For 2.5 hours the aircraft flew illegally over a piece of Soviet territory packed with sensitive military installations, and of course it was immediately assumed to be a spyplane.

At first a MiG-23 scrambled when the 747 was passing over the Kamchatka Peninsula but failed to find the target. Later, two Su-15TMs intercepted the huge airliner, trying to contact the crew by radio and by means of the usual signals and ordering it to land. Since the intruder ignored all calls and stubbornly pressed on towards the border, orders were given to destroy it. The flight leader, Maj. Ghennadiy Osipovich, fired two R-98 missiles which found their mark; the burning jumbo jet plunged into the Sea of Japan off Moneron Island, killing all 269 aboard. The incident provoked a huge public outcry and a massive anti-Soviet campaign led by the USA.

The reader may be interested to know how this mission proceeded. The following is a transcript of the radio exchange between the Soviet PVO command centre (CC, callsign *Deputat*, 'Deputy') and pilot (P) Maj. Osipovich (callsign '805').

P: *Deputat, this is 805, heading 45°, climbing to eight (8,000 m/ 26,250 ft – Auth.).*
CC: *805, roger, stay on this heading.*
P: *Copy that.*
05:56 CC: *805, Deputat here, the target is 5 [degrees] to port, range 130 [km/80.75 miles]. Target follows a heading of 240°, 5 [degrees] to port, range 120 [km/74.5 miles].*
05:58 CC: *805, Deputat here, target straight ahead, range 70 [km/43.5 miles], flight level 10,000 [m/32,800 ft].*
06:02 P: *Target in sight, flying at 8,000.*
06:03 CC: *Roger, target straight ahead, range 12-15 km [7.5-9.3 miles].*
06:04 CC: *805, Deputat here, target coded hostile, to be destroyed in the event of an incursion.* (This is the order to fire – Auth.). *Activate the special system* (ie, weapons control system – Auth.).
P: *Roger, wilco.*
06:10 CC: *805, can you identify the aircraft type?* (This query and the response are especially noteworthy– Auth.).
P: *Of course no.*
06:11 CC: *805, interrogate the target* (by means of IFF – Auth.).
P: *Roger.*
06:13 P: *No ['friendly'] response.*
CC: *Roger, activate the special system.*
P: *System active.*
CC: *805, Deputat here, keep an eye on the target heading.*
P: *Roger, so far it maintains the same heading.*
06:14 CC: *805, Deputat here, get ready to fire, get ready.*
P: *Roger. I'll need to use afterburners.*
CC: *What's your fuel status?*
P: *I got 2,700 [kg/5,950 lb].*
CC: *Engage afterburners when ordered.*
06:15 CC: *805, do you have a good lock-on?*
P: *I got a stable lock-on.*
06:17 CC: *Do you see the adversary?*
P: *Target in sight.*
CC: *Roger. Destroy!*
P: *Repeat that, please!*
06:18 CC: *805, the target has violated the state border. Destroy the target!*
P: *Roger, wilco.*
CC: *Does the target have the navigation lights on?*
P: *Affirmative, the nav lights are switched on... a flashing beacon is on* (the anti-collision light – Auth.).
CC: *Roger.*
06:19 CC: *805, flash your nav lights.*
CC: *805, flash your nav lights briefly.*
CC: *805, force the target to land on our airfield!* (This command overrules the one to destroy the target – Auth.).
P: *...I got my missile launch indicators on!*
CC: *805...?*
P: *805 here.*
06:20 CC: *805, fire a warning shot! Give a cannon blast!*
P: *...I have to move in closer. I'm cancelling lock-on, moving in.*
CC: *Give a cannon blast!*
CC: *805! Comply!*
P: *Lock-on cancelled, I'm firing the cannons.*
CC: *Have you fired, 805?*
P: *Affirmative.*
CC: *Do you see the target?*
06:21 P: *Yes, I'm closing in, moving closer.*
CC: *Roger.*
P: *The target is descending; I'm within 2 km [1.25 miles] or so.*
CC: *Is the target descending?*
P: *Negative, it is still flying at 10,000.*
AP: (another pilot, callsign '163'): *I see you both, about 10-15 km [6.2-9.3 miles] out.*
P: *The target is decelerating. I'm passing... passing... I'm ahead now.*

	CC:	*Roger, 805. Reduce speed, 805.*
06:22	CC:	*Flash your nav lights.*
	P:	*Roger, wilco. Speeding up now.*
	CC:	*Has the target increased speed?*
	P:	*Negative, it is reducing speed.*
	CC:	*805, open fire!*
	P:	*I cannot! You should have told me earlier… I'm off the target's wing now.*
	CC:	*Roger. Move into attack position if you can.*
	P:	*I'll have to fall behind now.*
	CC:	*Report relative target position, 805.*
	P:	*Repeat that, please!*
	CC:	*What is the target's altitude?*
	P:	*10,000.*
	CC:	*What's the relative target position? Relative to you?*
	P:	*Relative target position? Let me see… 70° to port.*
	CC:	*Roger.*
06:20	CC:	*805, try to destroy the target with cannon fire.*
	P:	*I'm falling behind, I'll try missiles.*
	CC:	*Roger.*
	CC:	*805, approach the target, destroy the target!*
	P:	*Roger, wilco. I got a good lock-on.*
06:20	CC:	*805, are you closing in?*
	P:	*Target ahead, I have lock-on, range 8 [km/5 miles].*
	CC:	*Engage afterburners!*
	P:	*'Burners on.*
	CC:	*Fire!*
	P:	*Missiles fired. Target destroyed.*
	CC:	*Break off the attack, turn right on heading 75°.*
	P:	*Attack interrupted.*

As the reader has probably realised, neither the officer at the PVO command centre nor the pilot were able to identify the aircraft they attacked because the incident took place at night. (However, even in poor lighting conditions the Boeing 747 is easy to identify by its unmistakable humpbacked silhouette.)

Even today it is not clear what the KAL jumbo was doing for 2.5 hours in restricted airspace. Was it on a premeditated spy mission, as the Soviet government claimed, or was the incursion a result of a grave navigation error? There are several possible explanations and facts to support both theories; however, this is a major topic which warrants a book in itself.

Of course, such incidents involving civil airliners do not speak volumes for the Su-15's virtues as an interceptor, being only pages in its service career. Still, according to Soviet fighter pilots' recollections, the reconnaissance aircraft of the 'potential adversary' took pains to avoid coming within the Su-15's reach. A notable exception is the Lockheed SR-71A Blackbird spyplane capable of Mach 3 flight; the MiG-25 was the only Soviet aircraft which was a match for the Blackbird. Of course, the presence of these very different interceptors in the PVO inventory (not counting the Tupolev Tu-128 heavy interceptor, the MiG-23P etc.) increased the overall efficiency of the nation's air defences. The potent MiG-25P was a complicated aircraft to build due to its welded steel airframe and its operations were hampered by the scarcity of its R15B-300 engines which were also used by the many reconnaissance/strike versions of the *Foxbat*; conversely, the less capable Su-15 was easy to build and well adapted for mass production as far as airframe, powerplant and equipment were concerned.

A major problem which the Soviet PVO had to deal with was the large number of drifting reconnaissance balloons launched from Western Europe. Frequently, despite all efforts to destroy it, such a balloon would pass over the entire country. Quite apart from their reconnaissance mission, the dastardly balloons presented a serious danger of collision for civil and military aircraft alike. Supersonic interceptors had limited success in combating reconnaissance balloons, primarily because the target usually had a very small radar cross-section; the aircraft's radar could only detect them at close range, which left very little time for an attack.

Su-15 pilots started their score in the autumn of 1974 when the first balloons were shot down. On 17th October PVO radar pickets detected Yet Another Evil Balloon drifting at 13,000 m (42,650 ft) over the Black Sea and about to enter Soviet airspace. Three 62nd IAP Su-15s took off from Bel'bek AB, making consecutive firing passes at the target; the last of the three managed to shoot off the balloon's reconnaissance systems pod with an R-98T missile. By far the greatest number of such sorties was flown in 1975 – and it was the most successful year as well, 13 out of 16 balloons being destroyed, including five downed by Su-15s.

It should be noted that most of the intruders the Su-15 had to deal with were anything but the typical targets it had been designed to intercept. As often as not the target was a light aircraft which was no easy target for a supersonic interceptor due to the huge difference in airspeeds. Moreover, the intruding light aircraft usually flew at ultra-low level where the interceptor's radar could not get a lock-on; this meant the target had to be located visually, and the view from the Su-15's cockpit left a lot to be desired. This was when accurate guidance by GCI centres proved crucial.

One of the first such incidents occurred on 21st June 1973. At 08:36 local time a radar picket of the Baku PVO District detected a target over Iranian territory 300 km (186 miles) south-east of Baku, moving towards the Soviet border at 2,000 m (6,560 ft). Five minutes later a Su-15 took off to ward off the potential intruder; it was soon joined by two more Su-15s of the 976th IAP which was temporarily deployed at Nasosnaya AB near Baku due to runway resurfacing work at their own base and a quartet of MiG-17PFUs of the 82nd IAP home-based at Nasosnaya AB.

The intruder crossed the Soviet border at 08:59 near the so-called Imishli Bulge 170 km (105 miles) south-west of Baku, descending to 200 m (660 ft) to avoid detection by radar. This complicated things considerably for the Su-15 pilots; nevertheless, at 09:09 the aircraft, a twin-engined Rockwell Aero Commander, was detected and hemmed in, making an involuntary landing at Nasosnaya AB 27 minutes later. It transpired that the pilot and the sole passenger were heading from Tabriz to the small borderside town of Parsaabad but had lost their way in the mountains. Well, well…

Conversely, it was no success story on 25th July 1976 when a 'visiting' Cessna 150 Aerobat got away. At 19:13 the low-flying intruder was visually detected by border guards troops on the ground, as the PVO radar pickets had missed it. At 19:27 a 431st IAP Su-15TM piloted by Capt. Vdovin took off from Afrikanda AB. Nevertheless, the Cessna insolently landed at the PVO reserve airfield at Alakurtti which was conveniently close at hand, the crew refuelled the aircraft, using a spare can of petrol, and continued on their eastward quest unhindered.

Approximately at 19:50 the GCI centre directed the Su-15 towards the intruder (which had not avoided detection altogether). Due to poor weather Vdovin was forced to fly below the clouds; still, he managed to spot the Cessna but then lost sight of it and could not regain contact. Two more Su-15TMs and a UTI-MiG-15 trainer (!) were never even directed towards the target. Thus the Finnish-registered Cessna flew on for another 300 km into the depths of the Karelian ASSR but then came to grief, flipping over on its back during a forced landing in a clearing in the woods. Soon afterwards the local residents found the crew and made a 'citizen's arrest'; the Finns claimed they had 'lost their bearings'.

A huge scandal erupted within the PVO system. The PVO C-in-C issued an order requiring that the pilots' gunnery training be stepped up; also, to ensure interception of low- and slow-flying targets like this one the QRA flights of Su-15 units were to include an

Above: In the mid-1980s a number of Su-15s was transferred from the IA PVO to the tactical aviation arm (FA), gaining a two-tone camouflage. This late-production Su-15TM appears to have a low-visibility tactical code, '24 White outline'.

aircraft armed with UPK-23-250 cannon pods by all means. As a result, from 1970 onwards the aircraft in a QRA flight were armed differently (for example, the flight leader carried two missiles (an R-98TM and an R-98RM) and two drop tanks while the wingman had the same complement of AAMs plus two cannon pods.

The mid-1980s saw a dramatic increase in the requirements which modern interceptors had to meet; new long-range AAMs and more capable aircraft to carry them were developed. Thus the Su-15 was relegated to second place in the PVO inventory, making way for such aircraft as the world-famous Su-27. Some of the Su-15s were transferred to the Soviet Air Force's tactical arm (FA), exchanging their natural metal finish for a green/brown tactical camouflage. However, the Su-15 was obviously no good as a strike aircraft, since it lacked the appropriate targeting equipment; actual operations soon confirmed this and the type did not gain wide use with the FA.

Like its precursors, the Su-15 was never exported; however, it did see overseas deployment. The 54th GvIAP deployed to Poland about once in every two years to practice operations from stretches of highway used as tactical strips; nothing of the sort existed in the USSR. For instance, in the summer of 1975 a squadron of the 54th GvIAP (by then equipped with the Su-15TM) was on temporary deployment at Słupsk.

The demise of the Soviet Union brought an end to the Su-15's service career in Russia. Even aircraft with plenty of airframe life remaining were struck off charge and scrapped in keeping with the Conventional Forces in Europe (CFE) limitation treaty. The Ukraine hung on to its Su-15s a little longer; the last of the type remained in service with the 636th IAP at Kramatorsk and the 62nd IAP at Bel'bek until 1996.

The Su-15 logically completed the line of Sukhoi's delta-winged interceptors that started with the Su-9, and its withdrawal was a bit hasty since the aircraft still had development and upgrade potential. It may have benefited from the installation of a new radar with 'look-down/shoot-down' capability, for instance. As a result, in 1976 the PVO fighter units started converting en masse to the MiG-23M which had this capability. This aircraft was, in turn, succeeded by the MiG-31M and the Su-27P representing a new generation of interceptor technology.

By way of conclusion, a curious fact deserves mention. In 1995 the Russian legal machine investigated the attempted sale of several Su-15s to private museums abroad. Several high-ranking Air Force officers were arrested and brought to trial; one of them, former IA PVO Chief of Engineering Maj. Gen. V. D. Ishootko, managed to escape from a hospital where he was undergoing treatment at the time of the investigation and flee to the Ukraine. When reporters from the Russian independent TV channel NTV tracked him down in June 1998, Ishootko stated that he had requested political asylum from the Ukrainian government! He added that the Su-15 had become a popular collector's item in the West due to the tragic shootdown of the Korean 747.

A few Su-15s are preserved in Russian museums. The Soviet Air Force Museum (now Central Russian Air Force Museum) in Monino has the T-58L development aircraft. The PVO Museum in Rzhev has a Su-15UM coded '46 Red'. The base museum at Savostleyka includes Su-15 '71 Red', Su-15TM '34 Red' and Su-15UM '30 Red'. Su-15TM '39 Red' (c/n 1209) is preserved at the Armed Forces Museum in Moscow; another example coded '11 Yellow' (c/n 1029) is on display in the Great Patriotic War Museum. Su-15 '01 Red' (c/n 0015301), Su-15s '42 Red' (c/n 0615342) and '85 Red' (c/n 0815344), Su-15T '37 Red' (c/n 0105) and Su-15UT '50 Red' (c/n 1015310) are on display at the now-defunct museum at Moscow-Khodynka. A further Su-15 resides in the PVO Museum in Nemchinovka immediately west of the Moscow city limits. Su-15s coded '45 Red' and '51 Red' are teaching aids at the Samara State Aviation University. Su-15TM '11 Red' and Su-15UM '91 Red' are preserved at a military memorial in Magadan. About half a dozen more Su-15s survive as museum exhibits and instructional airframes in the Ukraine.

An atmospheric nighttime shot of the flight line at an FA unit equipped with Su-15TMs, with '58 White outline' and '60 White outline' foremost; the fourth aircraft in the row is a Su-15UM.

Chapter 7

A Close Look at the Sukhoi Deltas

The Su-9/Su-11 in Detail

The following brief structural description applies to the standard Su-9. The differing design features of the Su-11 are indicated as appropriate.

Type: Single-engined single-seat supersonic interceptor designed for day and night operation in VMC and IMC. The all-metal airframe structure is made mostly of D-16 duralumin and V95 aluminium alloy; some highly stressed structural components are made of 30KhGSA and 30KhGSNA grade high-strength steel.

Fuselage: Semi-monocoque riveted stressed-skin structure of circular cross-section with 61 frames (66 on the Su-11), five longerons and 25 stringers. The fuselage diameter is 1.55 m (5 ft 1 in) between frames 15-28, increasing to a maximum of 1.634 m (5 ft 4⅜ in) on the rear fuselage.

Structurally the fuselage consists of two sections: forward (section F-1) and rear (section F-2), with a break point at frames 28/29 which allows the rear fuselage to be detached for engine maintenance and removal; the two fuselage sections are held together by bolts.

The *forward fuselage* is built in three portions, with manufacturing joints at frames 4 and 9. The annular forward portion incorporates an axisymmetrical circular air intake with a movable two-shock centrebody (shock cone); the latter is attached to a vertical splitter which divides the air intake into two elliptical-section air ducts passing along the fuselage sides. Four auxiliary blow-in doors are located on the forward fuselage sides ahead of the cockpit, opening as required to prevent engine surge at high rpm; the doors and shock cone are controlled by the ESUV-1 electrohydraulic air intake control system on the Su-9 or the ESUV-1S on the Su-11.

The portion between frames 4-9 incorporates the pressurised cockpit flanked by the inlet ducts and the nosewheel well located under it; the ducts merge again aft of the cockpit at frame 23. Neither the forward portion nor the centre portion of Section F-1 feature any stringers. The cockpit is enclosed by a two-piece bubble canopy. The fixed windscreen features two curved triangular Perspex sidelights and an elliptical optically-flat pane of bulletproof silicate glass; the aft-sliding canopy with blown Perspex glazing moves on guide rails and can be jettisoned manually or pyrotechnically in an emergency. The cockpit features an ejection seat (moving along guide rails attached to the rear pressure bulkhead), an instrument panel and side control consoles.

An avionics/equipment bay is located immediately aft of the cockpit, followed by the Nos 1 and 2 fuel tanks. Both forward fuselage fuel tanks were originally bag-type tanks (fuel cells) on early Su-9 production batches; later the No.1 fuel cell was replaced by an integral tank. Concurrently with this change the sloping fuselage frames Nos 17-19, which were originally intended to serve as guides for the wing cannons' ammunition belts, were replaced by ordinary ones set at right angles to the fuselage waterline.

Above: The forward fuselage of Su-9 '68 Red' (c/n 0615308) at the Soviet Air Force Museum in Monino, photographed in the days when the aircraft was still indoors. Note the auxiliary blow-in doors ahead of the windscreen and the closely spaced fuselage frames in the extreme nose.

The forward fuselage of Su-11 '14 Blue' (c/n 0115307) at Monino, again showing the auxiliary blow-in doors. Note the triple aerials of the SRO-2M Khrom-Nikel' IFF transponder ahead of the nosewheel well.

Above: The fuselage is divided into two major portions by a break point located just aft of the square air outlet grilles on the sides of the centre fuselage.

Above: The rear fuselage incorporates airbrakes and features numerous cooling air scoops. This view also illustrates the wing-mounted missile pylons with covers to prevent injury to ground personnel.

The centre fuselage incorporates two splayed pylons for drop tanks. Note the gun blast plates attached by multiple rows of rivets on this early-production Su-9.

The engine bay starts at frame 23, continuing aft to the fuselage break point. This part of the fuselage incorporates numerous removable access panels and cooling air intakes located on the underside.

The Su-9U trainer features a longer fuselage nose incorporating tandem cockpits for the trainee and instructor. These are enclosed by a common four-piece canopy featuring individual upward-hinged sections over the two seats. The provision of the instructor's cockpit required the No.1 fuel tank to be deleted.

Fuselage mainframes 15, 21, 25 and 28 incorporate wing attachment fittings. On the Su-11 the forward fuselage features two prominent wiring conduits (port and starboard) located on the upper fuselage sides.

The *rear fuselage* is a one-piece structure; the greater part of its internal volume is occupied by the engine's extension jetpipe. A titanium firewall is provided at frame 31. The rear fuselage incorporates the ventrally located No.3 fuel cell (or the No.3 integral fuel tank on the Su-11) and brake parachute bay. Four airbrakes with an area of 0.33 m^2 (3.54 ft^2) each are incorporated in a cruciform arrangement ahead of the tail unit; the airbrakes are electrohydraulically actuated and have a maximum deflection of 50°. Fuselage mainframes 38, 42 and 43 serve as attachment points for the fin and the horizontal tail.

Wings: Cantilever mid-wing monoplane with delta wings. Leading-edge sweep 60°, anhedral 2° from roots, incidence 0°.

The wings are of two-spar stressed-skin construction; they are one-piece structures joined to the fuselage at frames 15, 21, 25 and 28. Each wing has 14 ribs, 25 rib caps and three transverse beams (auxiliary spars) which, together with the front and rear spars, form five bays: the leading edge, forward bay, mainwheel well, rear bay and trailing edge. Early-production Su-9s had provisions for installing cannons in the forward bays; on later aircraft these bays were transformed into integral fuel tanks. The mainwheel wells are contained by the Nos 1 and 2 auxiliary spars. The space between the Nos 2 and 3 auxiliary spars is occupied by the rear integral wing tanks whose skin panels are stamped integrally with the ribs and stringers; ordinary sheet metal skins are used elsewhere.

The wings have one-piece Fowler flaps terminating at approximately half-span, with ailerons outboard of these. The flaps are hydraulically actuated; late-production aircraft feature a pneumatic emergency flap extension system. The ailerons are both aerodynamically balanced and mass-balanced. There are two pylons equipped with missile launch rails under each wing (one pylon on the Su-11).

Tail unit: Conventional swept cantilever tail surfaces of riveted stressed-skin construction. The *vertical tail* comprises a one-piece fin and an inset rudder. The fin is a single-spar structure with a rear auxiliary spar (internal brace), stringers and 16 ribs; it features a root fillet which is built integrally with the fuselage and consists of two portions divided by the fuselage break point at frame 28. The glassfibre fin cap incorporates a wire mesh antenna for the communications radio. The mass-balanced rudder is a single-spar structure.

The mid-set *horizontal tail* consists of all-movable slab stabilisers (stabilators) rotating on axles set at 48°30' to the fuselage axis; incidence in neutral position is –2°. Each stabilator is a single-spar structure with front and rear false spars, stringers and ribs. The stabilators feature anti-flutter weights projecting beyond the leading edge at the tips.

Landing gear: Hydraulically retractable tricycle type, with single wheel on each unit; the nose unit retracts forward, the main units inward into the wing roots. All three landing gear struts have levered suspension and oleo-pneumatic shock absorbers; the nose unit is equipped with a shimmy damper. The nose unit is castoring; steering on the ground is by differential braking.

On early-production Su-9s the nose unit had a 570 x 140 mm (22.4 x 5.5 in) K-283 non-braking wheel, while the main units had 800 x 200 mm (31.5 x 7.87 in) KT-50U brake wheels (KT = *koleso tormoznoye* – brake wheel); later production batches had a KT-38A brake-equipped nosewheel and KT-89 mainwheels of identical dimensions to the earlier models. The Su-11 features a 570 x 140 mm KT-100 nosewheel (later replaced by a 600 x 155 mm (23.6 x 6.1 in) KT-104 unit) and 880 x 230 mm (34.6 x 9.0 in) KT-69/4 mainwheels. On the Su-9, tyre pressure is 9 kg/cm² (128.5 psi) for the nosewheel and 12 kg/cm² (171.4 psi) for the mainwheels; on the Su-11 it is 10 kg/cm² (142.85 psi) and 13 kg/cm² (185.7 psi) respectively.

The nosewheel well is closed by twin lateral doors, the mainwheel wells by triple doors (one segment is hinged to the front spar, one to the root rib and a third segment attached to the oleo leg). All doors remain open when the gear is down.

The mainwheels are equipped with cerametallic disc brakes; the nosewheel has an expander-tube brake. To shorten the landing run the Su-9 is equipped with a PT-7 or PTZ-7B ribbon-type brake parachute (*parashoot tormoznoy*) housed in a special bay on the aft fuselage underside.

Powerplant: One Lyul'ka AL-7F-1 axial-flow afterburning turbojet rated at 6,240 kgp (13,760 lbst) at full military power and 9,200 kgp (20,280 lbst) in full afterburner. Late-production

Above: The starboard wing of a Su-9. Note that the wingtips are slightly cropped and that the aerodynamically balanced ailerons are tapered.

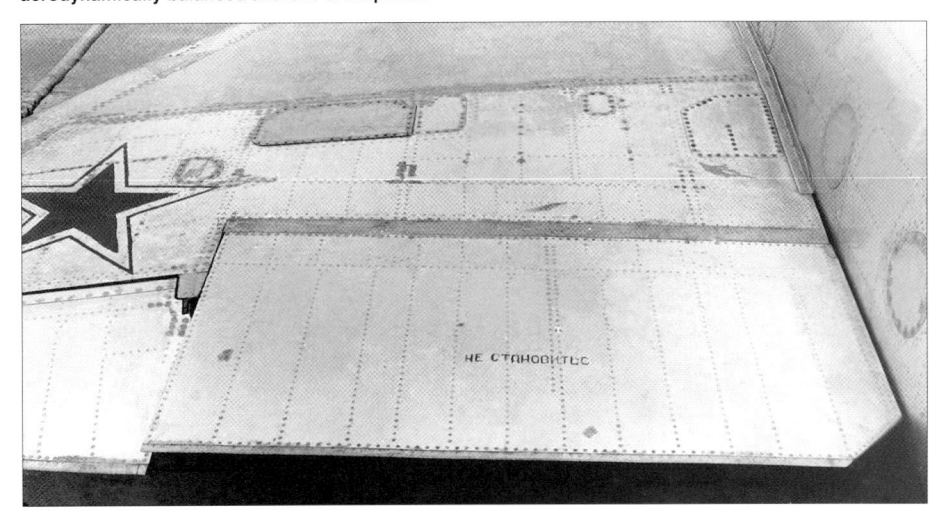

Above: Close-up of the port flap. The stencil reads *Ne stanovitsa* (No step).

The tail unit of a Su-9. This view shows the dielectric fin cap incorporating a communications radio antenna, the flush DME antenna and the rudder actuator access panel at the base of the fin.

Above: The port main landing gear unit of a Su-9. The angled outer segment of the main gear doors closes the cutout for the retraction

aircraft feature an AL-7F-1-100, AL-7F-1-150 or AL-7F-1-200 uprated to 6,800 kgp (14,990 lbst) dry and 9,600 kgp (21,160 lbst) reheat, with a designated service life increased to 100, 150 or 200 hours respectively. The engine was produced by the 'Salyut' Moscow Engine Production Enterprise and the Rybinsk Engine Factory.

The AL-7F is a single-spool turbojet featuring an intake assembly with a fixed spinner and 12 radial struts, a nine-stage compressor with a supersonic first stage, an annular combustion chamber with 18 vortex-type flame tubes, a two-stage turbine and an afterburner with a variable nozzle. The compressor has bleed valves at the fifth and seventh stages. The straight-through afterburner features inner and outer ducts, annular flame holder grids and anti-vibration shielding; the two-position convergent-divergent nozzle has 24 petals.

Engine accessories are driven via a ventral accessory gearbox whose power take-off shaft is located aft of the compressor. Starting is by means of a TS-19A or TS-20 turbostarter – a small gas turbine engine driving the spool directly via a clutch (the term 'jet fuel starter' is not applicable, since the starter runs on aviation gasoline); there are two ignition units with centrifugal fuel spray nozzles and spark plugs in two of the combustion chamber's flame tubes.

The AL-7F has an all-mode hydromechanical fuel control unit (FCU) and a closed-type lubrication system with a fuel/oil heat exchanger.

Engine pressure ratio (EPR) 9.1 (AL-7F-1); mass flow at take-off rating 114 kg/sec (251 lb/sec), normal turbine temperature at take-off rating 1,133° K, maximum turbine temperature 1,200° K. Specific fuel consumption (SFC) 2.0 kg/kgp·h (lb/lbst·h) in full afterburner and 0.91 kg/kgp·h in cruise mode; some sources state 2.3 and 0.96 kg/kgp·h respectively.

Length overall (including afterburner) 6,630 mm (21 ft 9 in), casing diameter 1,250 mm (4 ft 1¼ in). Dry weight 2,010 kg (4,430 lb; some sources state 2,050 kg/4,520 lb); weight of fully dressed engine 2,325 kg (5,125 lb). The final production batches had a 250-hour service life.

The Su-11 is powered by an AL-7F-2 with a maximum afterburner rating of 10,100 kgp (22,270 lb st). This version features a modified compressor with titanium first/second stage discs and improved eighth and ninth stages to increase the EPR to 9.3, a modified second turbine stage, a larger-diameter afterburner, a modified control system with turbine temperature and speed limiters, a TS-20B turbostarter and a new oil pump. Mass flow at take-off rating 115 kg/sec (253.5 lb/sec), maximum turbine temperature 1,200° K. SFC 2.0 (some sources say 2.25) kg/kgp·h in full afterburner and 0.89 kg/kgp·h in cruise mode. Length overall (including afterburner) 6,650 mm (21 ft 9⅞ in), maximum diameter 1,300 mm (4 ft 3⅛ in); dry weight 2,100 kg (4,630 lb). The service life is increased to 300 hours.

This view illustrates the inner faces of the mainwheels and the levered suspension of the main gear units. Note how the fins of the drop tanks overlap.

The Lyul'ka AL-7F-1 engine complete with extension jetpipe and afterburner. The accessories are located ventrally.

The engine breathes through a circular supersonic air intake with a movable shock cone and auxiliary inlet doors, both controlled by the ESUV-1 system (or ESUV-1S on the Su-11). The shock cone is in the fully aft position in subsonic flight, gradually moving forward as the Mach number increases to provide the optimum position of the shock waves ensuring stall-free engine operation throughout the speed envelope. The Su-11 features a pneumatic foreign object damage prevention system, engine bleed air being routed via a pipeline on the port side of the lower fuselage to twin ejector nozzles under the intake lip to create increased pressure, thus preventing debris ingestion; the system was not used in operational service.

Control system: Powered controls with BU-49 irreversible hydraulic actuators throughout. Control inputs are transmitted to the aileron and stabilator actuators by push-pull rods, control cranks and levers; a combined control linkage utilising both rods and cables is used in the rudder control circuit. Spring-loaded artificial-feel units are provided in all three control circuits; the rudder control circuit features two artificial-feel units – one for take-off/landing (disabled by landing gear extension) and the other for cruise flight. The stabilator control circuit includes an ARZ-1 stick force limiter (*avtomaht regooleerovaniya zagroozki*) which adjusts the stick forces depending on the flight mode, a differential mechanism altering the stick/tailplane movement ratio and a trim mechanism. The aileron control circuit features spring-loaded control rods for emergency manual control in the event that one of the actuators fails. An AP-106M yaw damper is provided.

The Su-11 features an AP-28Zh-1B autopilot with RA-16 servos in all three control circuits and a D-3K-110 yaw damper replacing the predecessor's AP-106M.

Fuel system: On early Su-9s internal fuel was carried in three fuel cells (two in the forward fuselage and one in the rear fuselage) and two integral tanks in the wing torsion box (aft of the mainwheel wells) holding a total of 3,060 litres (673.2 Imp gal). The replacement of the No.1 fuel cell by an integral tank and the provision of two more integral tanks in the wings on late-production aircraft increased the total capacity to 3,780 litres (831.6 Imp gal). There are provisions for carrying two 600-litre (132 Imp gal) drop tanks on pylons under the centre fuselage. The Su-11 has an internal fuel capacity of 4,195 litres (922.9 Imp gal) and carries larger 720-litre (158.4 Imp gal) drop tanks. Fuel grades used are Russian T-1, TS-1 or RT jet fuel for the engine and B-70 aviation gasoline for the turbostarter.

Electrics: Main electric power provided by a 12-kW GS-12T DC generator and an SGO-8 single-phase AC generator. Backup DC power provided by a 12SAM-25 (28 V, 25 A h) silver-zinc battery (or, on the Su-11, a 12-ASAM-23 battery) in the avionics/equipment bay. Stable frequency AC power for some systems is supplied by four PO-4500 single-phase AC converters (*preobrazovahtel' odnofahznyy*), one PT-1000Ts three-phase AC converter and one PT-500Ts three-phase AC converter (*preobrazovahtel' tryokhfahznyy*). The Su-11 features an additional PT-1200E AC converter for powering the weapons control system.

Early-production Su-9s had ultra-violet lights for the instrument panel to make the dials glow in the dark, later replaced by red cockpit lighting. The exterior lighting included port (red) and starboard (green) BANO-45 navigation lights at the wingtips and a white tail navigation light on the fin trailing edge. The Su-9 had an FR-100 taxi light on the nose gear strut and retractable LFSV-45 landing lights in the wing roots; on the Su-11 these three lights were replaced by two retractable PRF-4M landing/taxi lights.

Hydraulics: Three separate hydraulic systems, each with its own NP-26/1 engine-driven plunger-type pump (*nasos ploonzhernyy*). The primary system operates the landing gear, flaps, airbrakes, air intake shock cone and auxiliary blow-in doors; it also performs automatic wheel braking during landing gear retraction. The two actuator supply systems (main and back-up) power exclusively the aileron, rudder and tailplane actuators; in addition to the engine-driven pump, the back-up system features an NP-27 autonomous emergency pump ensuring that the system remains operational (and hence the aircraft remains controllable) in the event of an engine failure. All systems use AMG-10 oil-type hydraulic fluid (*aviatsionnoye mahslo ghidravlicheskoye*); nominal pressure 210 kg/cm² (3,000 psi).

The Su-11 differs in having an NP34-1T pump in the primary hydraulic system and NP-26/3 pumps in the actuator supply systems, plus an NS-3 emergency pump in the back-up system. The primary system also operates the radar scanner drive and autopilot servos.

air pressure is governed by an ARD-57V automatic pressure regulator. The air is fed to the cockpit via nozzles under the canopy transparencies, demisting them at the same time.

Oxygen system and high-altitude equipment: A KKO-2 or, in the case of the Su-11, KKO-3 oxygen equipment set (*komplekt kislorodnovo oboroodovaniya*) is provided for high-altitude operations; it also ensures pilot survival in the event of decompression. The KKO-2 comprises gaseous oxygen bottles, pressure reduction gear, a KM-3-OM oxygen mask, a KP-34 breathing apparatus (*kislorodnyy pribor*) for normal operation and a KP-27M breathing apparatus used in the event of an ejection at high altitude.

The pilot is equipped with a VKK-3M pressure suit (*vysotnyy kompenseeruyuschchiy kostyum* – altitude compensation suit) and a GSh-4M full-face pressure helmet (*ghermoshlem*) permitting safe ejection at high altitude. Su-11 pilots used the VKK-4M suit and GSh-4MS pressure helmet.

Fire suppression system: The hot zone of the engine is isolated from the airframe by a firewall at frame 31 and a heat shield around the engine. Fire extinguisher bottles charged with carbon dioxide are provided. System operation is manual; in the event of engine fire several flame sensors trigger a fire warning light in the cockpit and the pilot pushes a button, activating pyrotechnic valves and letting out the carbon dioxide into a manifold around the engine.

Armament: The Su-9 was originally armed with four RS-2-US semi-active radar homing (SARH) air-to-air missiles. These are carried on pylon-mounted APU-19 launch rails on the outboard hardpoints (Nos 1 and 4) and APU-20 launch rails on the inboard pair (Nos 2 and 3). Later the aircraft's weapons system was upgraded by the addition of R-55 infra-red homing AAMs, and the normal ordnance load came to consist of two RS-2-USs and two R-55s. The missiles can be fired singly, in pairs (with a 0.35-second interval) or all together with a 13-second interval between the pairs; the launch sequence in the latter case is 2-3-1-4.

The Su-11 is armed with two R-8M AAMs carried on PU-1-8 launch rails. Usually the complement includes one R-8MR SARH missile and one R-8MT IR-homing missile. The missiles can be fired singly or in a salvo with a 0.5-second interval.

Avionics and equipment
Navigation and piloting equipment: GIK-1 gyro-flux gate compass on early-production Su-9s, replaced by a KSI compass system on later aircraft. RSP-6 instrument landing system including an ARK-5 Amur automatic direction finder (ADF) with omnidirectional aerial

Two RS-2-US air-to-air missiles on the wing pylons of a Su-9. This view illustrates the lateral nozzles of the missile's rocket motor, a feature necessitated by the aft-mounted receiver antenna of the guidance system.

Pneumatic system: Two subsystems (main and emergency). The pneumatic system operates the wheel brakes and inflatable canopy perimeter seal; it is also responsible for emergency landing gear and flap extension in the event of hydraulics failure. The system is charged with nitrogen to 150 bars (2,140 psi). The Su-9 has three nitrogen bottles with a total capacity of 12 litres (2.64 Imp gal); the Su-11 features five bottles with a total capacity of 21 litres (4.62 Imp gal).

Air conditioning and pressurisation system: The cockpit is pressurised by air bled from the engine's fifth or seventh compressor stage, depending on rpm, to ensure proper working conditions for the pilot at high altitude. The canopy is sealed by an inflatable rubber hose running around the perimeter.

Cockpit air temperature is maintained automatically at +10-20°C (50-68°F) by a TRTVK-45M regulator; on the Su-11 the temperature limits are +16-20°C (+60-68°F). Cockpit

and loop aerial (replaced by the ARK-10 on the Su-11), an RV-UM low-range radio altimeter, an MRP-56P marker beacon receiver and SOD-57M distance measuring equipment (**stahn**tsiya oprede**len**iya **dahl**'nosti).

Communications equipment: RSIU-4V two-way VHF radio (replaced by the RSIU-5 on the Su-11) with a mesh-type antenna built into the dielectric fin cap. The aircraft has a data link receiver making up part of the Lazoor' (ARL-S) ground controlled intercept (GCI) system.

Weapons control system: The Su-9 is equipped with an RP-9U fire control radar whose radome is part of the air intake shock cone; aircraft upgraded to take the R-55 AAM have a modified RP-9UK radar. The weapons control system of the Su-11 is built around a similarly positioned RP-11 fire control radar. An AKS-5 gun camera buried in the starboard wing leading edge records missile launches, and a PAU-457 photo module records the target (and the destruction thereof) on the radar display.

Flight instrumentation: KUSI-2500 airspeed indicator, VDI-30 altimeter, AGI-1 artificial horizon (replaced by an AGD-1 on late-production aircraft), EUP-53 turn and bank indicator, AM-10 accelerometer (G load indicator), VAR-300 vertical speed indicator (variometr), GIK-1 gyro-flux gate compass indicator on early-production Su-9s replaced by a UKL-1 heading indicator (or a UKL-2 on the Su-11), KI-13 compass (Su-11 only),

An R-98R missile on the port wing pylon of a Su-11. The canard foreplanes are fixed, control being effected by the inset rudders. The missile carries the c/n ЕГ0050301085 (ie, YeG0050301085).

M-2,5 Mach meter, ARK-5 ADF indicator and AChKh clock.

Air data is provided by a pitot boom at the tip of the radome.

IFF equipment: SRZO-2M Kremniy-2M IFF transponder. The triple IFF aerials are located ventrally just ahead of the nosewheel well and the tail bumper.

Rescue system: Early-production Su-9s were equipped with a Sukhoi KS-1 ejection seat with an ejection speed limit of 850 km/h (528 mph); on later aircraft it was replaced by a more refined KS-2 which increased the maximum safe ejection speed to 1,000 km/h (620 mph). Minimum safe ejection altitude with these seat types is 150 m (490 ft) in level flight at speeds not less than 500 km/h (310 mph). The ultimate batches featured the improved KS-2A version. The Su-11 features a KS-3 ejection seat which expanded the operational envelope to a maximum speed of 1,100 km/h (683 mph) and a minimum safe ejection altitude of 30 m (100 ft).

Su-9 and Su-11 specifications

	Su-9	Su-9U	Su-11
Powerplant	AL-7F-1	AL-7F-1	AL-7F-2
Take-off rating, kgp (lbst):			
dry	6,800 (14,990)	6,800 (14,990)	6,800 (14,990)
reheat	9,600 (21,160)	9,600 (21,160)	10,100 (22,270)
Length (including pitot)	18.055 m (59 ft 2⅞ in)	18.655 m (61 ft 2½ in)	17.546 m (57 ft 6¾ in) *
Wing span	8.536 m (28 ft 0 in)	8.536 m (28 ft 0 in)	8.536 m (28 ft 0 in)
Height on ground	4.82 m (15 ft 9¾ in)	4.82 m (15 ft 9¾ in)	4.7 m (15 ft 5 in)
Wing area, m2 (sq.ft)	34.0 (365.5)	34.0 (365.5)	34.0 (365.5)
TOW, kg (lb):			
normal	11,442 (25,225)	11,773 (25,955)	12,674 (27,940)
maximum (with drop tanks)	12,512 (37,583)	12,863 (28,357)	13,986 (30,833)
Empty weight, kg (lb)	7,675 (16,920)	n.a.	8,562 (18,875)
Internal fuel load, kg (lb)	3,100 (6,835)	3,100 (6,835)	3,440 (7,580)
Top speed at 12,000 m (39,370 ft), km/h (mph):			
one-minute afterburner engagement	2,230 (1,385)	2,230 (1,385)	n.a.
prolonged afterburner engagement	2,120 (1,315) †	2,100 (1,304) ‡	2,340 (1,450)
Service ceiling, m (ft)	20,000 (65,620)	19,700 (64,630)	18,000 (59,050)
Range, km (miles):			
on internal fuel	1,350 (838)	1,130 (700)	1,260 (780)
with drop tanks	1,800 (1,118)	1,370 (850)	1,710 (1,060)
Take-off run, m (ft)	1,200 (3,940)	1,350 (4,430)	1,100-1,250 (3,600-4,100)
Landing run without brake parachute, m (ft)	1,150-1,250 (3,770-4,100)	1,200 (3,940)	1,000-1,200 (3,440-3,940)

* Less pitot/air intake shock cone in fully aft position
† Speed limited to Mach 2.1
‡ With two K-5M (RS-2-US) AAMs

Above: The forward fuselage of a Su-15 sans suffixe, *showing the forward avionics bay access panels. Note the bulges on the nose gear doors to accommodate the nosewheel.*

The Su-15 in Detail

Type: Twin-engined single-seat supersonic interceptor designed for day and night operation in VMC and IMC. The all-metal airframe structure is made mostly of D-16 duralumin, V95 and AK4-1 aluminium alloys; some highly stressed structural components are made of 30KhGSA, 30KhGSNA and 30KhGSL grade high-strength steel, while some components of the rear fuselage structure subjected to high temperatures are made of OT4-1 titanium alloy.

Fuselage: Semi-monocoque riveted stressed-skin structure. The fuselage cross-section changes from circular (in the forward fuselage) to almost rectangular (in the area of the air intakes) to elliptical with the longer axis horizontal (in the rear fuselage).

Structurally the fuselage consists of two sections: forward (Section F-1) and rear (Section F-2), with a break point at frames 34/35 which allows the rear fuselage to be detached for engine maintenance and removal; the two fuselage sections are held together by bolts.

The *forward fuselage* includes the radome, the pressurised cockpit, two avionics/equipment bays fore and aft of it, the nosewheel well under the cockpit, the air intake assemblies and inlet ducts, the engine bays and the fuel tank bays. The detachable dielectric radome has a metal attachment 'skirt'; it has a simple conical shape on the Su-15/Su-15T/Su-15U and an ogival shape on the Su-15TM/Su-15UT. The radome mounts the main pitot boom.

The forward avionics/equipment bay (frames 1-4) houses the radar set; some structural changes were made to this area on the Su-15T due to the installation of the new Taifoon radar. On single-seat versions the cockpit is contained by pressure bulkheads at frames 4 and 14A and is enclosed by a bubble canopy of similar design to that of the Su-9/Su-11. The Su-15UT trainer features a longer fuselage nose with tandem cockpits enclosed by a common canopy similar to that of the Su-9U. In contrast, the Su-15UM trainer does not have a fuselage stretch, being of identical dimensions to the single-seat Su-15TM on which it is based; the instructor's cockpit was incorporated by deleting part of the equipment.

The forward fuselage is area-ruled, starting at the cockpit section which is flanked by the air intakes. The lateral air intake trunks are rectangular-section structures blending gradually into the centre portion of the fuselage. The intakes have sharp lips and the leading edges are vertical in side view. To prevent boundary layer ingestion the intakes are set apart from the fuselage and feature boundary layer splitter plates; V-shaped fairings spilling the boundary layer connect the splitter plates to the fuselage. To improve performance at high angles of attack the air intake trunks are canted outward 2°30'. Each intake features a three-segment vertical airflow control ramp and a rectangular auxiliary blow-in door on the outer face; these are governed by the UVD-58M engine/intake control system.

The centre portion of Section F-1 (frames 14A-34) accommodates three integral fuel tanks, the inlet ducts and the engine bays (the latter are located between frames 28-34); the inlet ducts' cross-section gradually changes to circular at the engine compressor faces. On the first three T-58 prototypes this section was area-ruled between frames 14B and 28, with a maximum width of only 1.64 m (5 ft 4½ in). However, the 'waist' was eliminated on the first and third aircraft by riveting on false fuselage 'sidewalls' between frames 12 and 31. Afterwards the entire centre fuselage section of the T58D-1 was replaced with a new sub-assembly incorporating a lift engine bay with appropriate dorsal intake doors and ventral exhaust ports in the course of conversion into the experimental T-58VD. All subsequent examples had a constant-width centre fuselage section making it possible to increase the internal fuel capacity.

Fuselage mainframes 16, 21, 25, 28 and 29 incorporate wing attachment fittings. The rear end of Section F-1 mounts the fin fillet attached to the skin by special plates with anchor nuts.

The centre and rear fuselage of a Su-15 used as a ground instructional airframe; the rear fuselage is detached and many access panels are missing. Note the kinked wing leading edge in line with the pylon.

The *rear fuselage* incorporates the vertical tail attachment fittings, stabilator mounting axles and a detachable 'pen nib' fairing fitting between the engine nozzles; it accommodates the engine jetpipes and the tailplane actuators. The tail fairing is constructed of titanium and stainless steel sheet. Four airbrakes with a total area of 1.32 m² (14.19 ft²) are located between frames 35-38, opening in a cruciform arrangement; maximum deflection is 50°. Fuselage mainframes 38, 42 and 43 serve as attachment points for the fin and the all-moving horizontal tail.

Wings: Cantilever low-wing monoplane. The initial production version ('Su-15 *sans suffixe*') features simple delta wings similar to those of the Su-9/Su-11 with constant 60° leading-edge sweep, 2° anhedral and zero incidence. Aircraft from c/n 1115331 onwards (the Su-15T and Su-15TM) have increased-area wings featuring a leading-edge kink at 2.625 m (8 ft 7⅜ in) from the fuselage centreline, with 60° leading-edge sweep inboard and 45° outboard; the outer portions feature 7° negative camber.

The wings are of two-spar stressed-skin construction; they are one-piece structures joined to the fuselage at frames 16, 21, 25, 28 and 29. On the pure delta version each wing has 17 ribs, 28 rib caps and three auxiliary spars which, together with the front and rear spars, form five bays: the leading edge, forward bay, mainwheel well (between the Nos 1 and 2 auxiliary spars), rear bay and trailing edge. Later versions with a kinked leading edge have 18 ribs and 29 rib caps per wing.

As on the Su-9/Su-11, the rear bays between the Nos 2 and 3 auxiliary spars function as integral fuel tanks whose skin panels are stamped integrally with the ribs and stringers; ordinary sheet metal skins are used elsewhere.

The wings have one-piece blown flaps, with ailerons outboard of these. The flaps are hydraulically actuated, with pneumatic extension in an emergency. Flap settings are 15° for take-off and 25° for landing when the boundary layer control system (BLCS) is activated or 25°/45° with the BLCS inactive. The ailerons are aerodynamically balanced and mass-balanced, with a maximum deflection of ±18°30'. There are two pylons equipped with missile launch rails under each wing (one pylon on the Su-11).

Tail unit: Conventional tail surfaces of riveted stressed-skin construction; sweepback at quarter-chord 55° on all tail surfaces. The *vertical tail* comprises a one-piece fin and an inset rudder; it has greater area than the Su-9/Su-11's. On the first three prototypes the vertical tail was a stock Su-9 subassembly; in order to accommodate a brake parachute container at the base of the rudder the designers chose to simply insert a section at the base rather than redesign the fin. On the production model the structure was redesigned after all; the forward fin attachment fitting was moved forward from fuselage frame 38 to frame 35 and the fin spar was kinked to accomplish this.

The fin is a single-spar structure with a rear auxiliary spar (internal brace), front and rear false spars, stringers and ribs; it features a root fillet and a glassfibre tip fairing. The mass-balanced rudder is a single-spar structure; it is carried on three brackets.

The low-set cantilever *horizontal tail* mounted 110 mm (4⅜ in) below the fuselage waterline consists of differentially movable slab stabilisers (stabilators); anhedral 2°, incidence in neutral position –4°10'. Each stabilator is a single-spar structure with front and rear false spars, stringers and ribs, and sheet duralumin skins. The stabilators feature anti-flutter booms projecting beyond the leading edge at the tips; originally they were straight but then were angled 15° upwards on the production aircraft.

Landing gear: Hydraulically retractable tricycle type, with pneumatic extension in an emergency; the nose unit retracts forward, the main units inward into the wing roots. All three landing gear struts have levered suspension and oleo-pneumatic shock absorbers; the nose unit is equipped with a shimmy damper. The nose unit is castoring through ±60°; steering on the ground is by differential braking.

On the Su-15 *sans suffixe* the nose unit has a single 660 x 200 mm (26.0 x 7.87 in) KT-61/3 brake wheel. The Su-15T/Su-15TM

Above: The cockpit canopy of a Su-15. The stencil on the sliding portion reads Na stoyanke fonar' zachekhlyat' (Cover the canopy when aircraft is parked).

The port wing of a late-production Su-15 or a Su-15T/Su-15TM. This view shows the cranked leading edge with 60° leading-edge sweep inboard and 45° outboard. Note the wing fence, the pitot outboard of it and the increased-span tapered aileron.

feature a taller nose gear unit equipped with twin 62 x 18mm (24.4 x 7.0 in) KN-9 wheels (*koleso netormoznoye* – non-braking wheel) to cater for the new and heavier radar; the same twin-wheel unit is fitted to the radarless Su-15UM. The main units are equipped with 880 x 230 mm (34.6 x 9.0 in) KT-117 brake wheels on all versions.

The nosewheel well is closed by twin lateral doors, the mainwheel wells by triple doors (one segment is hinged to the front spar, one to the root rib and a third segment attached to the oleo leg). All doors remain open when the gear is down.

All brake wheels feature pneumatically actuated disc brakes with bi-metal and cera-metallic brake pads. A PT-15 ribbon-type brake parachute with an area of 25 m² (268.8 sq.ft) is housed in a fairing at the base of the fin.

Powerplant: The Su-15 and Su-15UT are powered by two Tumanskiy R-11F2S-300 or R-11F2SU-300 axial-flow afterburning turbojets rated at 3,900 kgp (8,600 lbst) at full military power and 6,175 kgp (13,610 lb st) in full afterburner. (Some sources give different figures – 4,200 kgp (9,260 lbst) dry and 6,120 kgp (13,490 lb st) reheat.)

The R-11F2S-300 is a two-spool turbojet featuring a three-stage low-pressure (LP) compressor, a three-stage high-pressure (HP) compressor, an annular combustion chamber, single-stage HP and LP turbines and an afterburner with an all-mode variable nozzle. The transonic-flow compressor has no inlet guide vanes, which facilitates replacement of the first LP stage in the event of damage; the second-stage blades have snubbers to prevent resonance vibrations. Bleed valves are provided.

Engine accessories are driven via a ventral accessory gearbox whose power take-off shaft is located aft of the HP compressor. The engine has a closed-type lubrication system. Starting is by means of a starter-generator, using DC power from on-board batteries or ground power.

Top: The starboard air intake of a Su-15. The stencils in between the black/yellow stripes on the boundary layer splitter plate read *Opasno, vozdookhozabornik* (Danger, air intake). Note the perforated forward segment of the hinged airflow control ramp and the auxiliary blow-in door just aft of the tactical code.

Centre: The vertical tail of a Su-15 in the PVO's technical school in Solntsevo, showing the dielectric fin cap, the flush SHORAN antenna below it and the rudder actuator access panel.

Bottom: The rear fuselage and tail unit of the Su-15TM prototype (c/n 0105) at Moscow-Khodynka. Regrettably the dielectic fin cap is missing. This view illustrates the slab stabilisers with upward-angled anti-flutter booms and the open brake parachute container.

The R-11F2SU-300 version features air bleed ducts for the boundary layer control system. The bleed rate does not exceed 2.5 kg/sec (5.5 lb/sec) and the BLCS operates with the engines running in dry thrust mode.

EPR at take-off rating 8.7; mass flow at take-off rating 66 kg/sec (145.5 lb/sec), maximum turbine temperature 1,175° K. SFC 2.37 kg/kgp h (lb/lbst h) in full afterburner and 0.93 kg/kgp h in cruise mode; the said sources state 2.2 and 0.94 kg/kgp h respectively. Length overall (including afterburner) 4,600 mm (15 ft 1 in), casing diameter 825 mm (2 ft 8½ in). Dry weight 1,088 kg (2,400 lb).

Aircraft from c/n 1115331 onwards (the Su-15T, Su-15TM and Su-15UM) are powered by two Tumanskiy R-13-300 axial-flow afterburning turbojets rated at 4,100 kgp (9,040 lbst) at full military power and 6,600 kgp (14,550 lb st) in full afterburner. This is likewise a two-spool turbojet with a three-stage LP compressor, a five-stage HP compressor, an annular combustion chamber, single-stage HP and LP turbines and an afterburner with an all-mode variable nozzle. The afterburner features annular/radial flame holders and a perforated heat shield.

EPR 9.15 at full military power and 9.25 in full afterburner; mass flow at take-off rating 66 kg/sec, max turbine temperature 1,223° K. SFC 2.25 kg/kgp h (lb/lbst h) in full afterburner and 0.96 kg/kgp h in cruise mode. Length overall (including afterburner) 4,600 mm, casing diameter 907 mm (2 ft 11¾ in). Dry weight 1,134 kg (2,500 lb).

Control system: Powered controls with BU-49 or BU-220D irreversible hydraulic actuators. Control inputs are transmitted to the aileron and stabilator actuators via push-pull rods, control cranks and levers; in the rudder control circuit, cables are used up to frame 31, with rigid linkage further aft. Artificial-feel units are provided. The tailplane control circuit includes an ARZ-1 stick force limiter, a differential mechanism and a trim mechanism. There is no autopilot.

Fuel system: Internal fuel is carried in five integral tanks – three fuselage tanks (No.1, frames 14A-18; No.2, frames 18-21; No.3, frames 21-28) and two wing tanks located immediately aft of the mainwheel wells, as on the Su-9/Su-11. The total capacity is stated variously in different documents, the figures ranging from 8,675 to 8,860 litres (1,907 to 1,949 Imp gal). The distribution for the latter figure is as follows: No.1 fuselage tank, 2,350 litres (517 Imp gal); No.2 tank, 1,150 litres (253 Imp gal); No.3 tank, 2,740 litres (602.8 Imp gal); and 615 litres (135.3 Imp gal) in the wing tanks. There are two 'wet' hardpoints under the centre fuselage permitting carriage of two 600-litre (132 Imp gal) drop tanks.

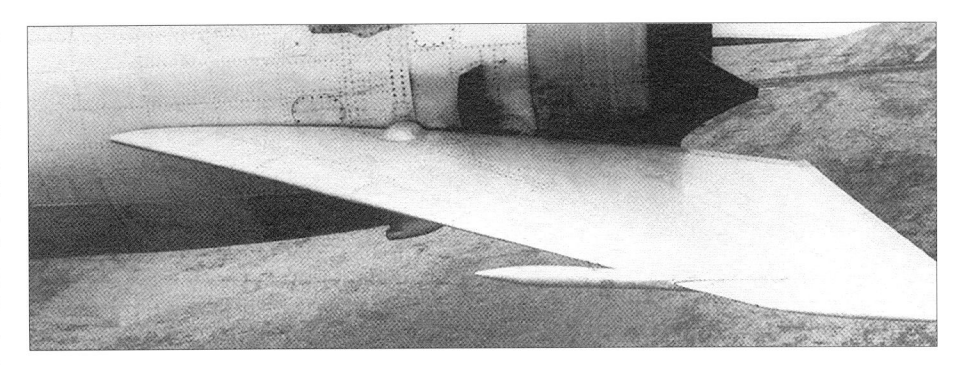

Above: Close-up of the experimental increased-area stabilators tested on Su-15TM '75 Blue' (c/n 0303). Note the straight (not angled) anti-flutter boom.

Electrics: Main 28.5 V DC power provided by two 12-kW GSR-ST-12000VT engine-driven starter-generators, with two 15-STsS-45A silver-zinc batteries providing 22.5 V DC as a back-up source. 115 V/400 Hz single-phase AC provided by two SGO-8TF engine-driven generators. Two ground power receptacles are provided on the port side of the fuselage. The electric system uses a single-wire layout.

The exterior lighting inluded port (red) and starboard (green) BANO-45 navigation lights at the wingtips, a KhS-39 white tail navigation light on the fin trailing edge near the top, identical KhS-39 lights on all three landing gear struts to confirm to the ground personnel that the gear is 'down and locked' during night operations, and two retractable PRF-4M landing/taxi lights in the wing roots.

The twin-wheel nose landing gear unit of the T-58L; the same design is used on the Su-15T/TM/UM.

99

engine failure. All systems use AMG-10 oil-type hydraulic fluid; nominal pressure 215 kg/cm^2 (3,070 psi).

Pneumatic system: The pneumatic system performs normal and emergency wheel braking, emergency landing gear and flap extension, and operates the inflatable canopy perimeter seal. The system is charged with compressed air to 200 bars (2,857 psi), featuring three 6-litre (1.32 Imp gal) air bottles. There is also a separate pneumatic system charged to 150 bars (2,140 psi) which operates the stabilising gyros of the R-98 (or R-8M) missiles' seeker heads.

Air conditioning and pressurisation system: Similar to that of the SU-9/Su-11. The cockpit is pressurised by engine bleed air; cockpit air pressure and temperature are maintained automatically. The canopy is sealed by an inflatable rubber hose running around the perimeter.

Oxygen system and high-altitude equipment: The oxygen equipment includes a breathing apparatus for normal operation and a separate breathing apparatus used in the event of an ejection.

For operations at altitudes up to 10,000 m (32,810 ft) and speeds up to 900 km/h (560 mph) the pilot is equipped with a KM-32 oxygen mask, a ZSh-3 'bone dome' flying helmet (*zaschchitnyy shlem*) and a VK-3 or VK-4 ventilated flying suit (*ventileeruyemy kostyum*). For missions involving supersonic flight the pilot wears a VKK-4M, VKK-6 or VKK-6P pressure suit and a GSh-4MS, GSh-6M or GSh-4MP full-face pressure helmet.

Above: Another view of the Su-15TM's nose landing gear unit with KN-9 non-braking wheels. Note the steering actuators above the torque link.

Hydraulics: Four separate hydraulic systems (two primary and two actuator supply systems), each with its own NP-34 or NP-26/1 engine-driven plunger-type pump. The No.1 primary system operates the landing gear, flaps, airbrakes, artificial-feel unit switch mechanism, the port air intake ramp and auxiliary blow-in door and the port engine's nozzle actuators. It also performs automatic wheel braking during landing gear retraction. The No.2 primary system powers the radar dish drive, the starboard air intake ramp/auxiliary blow-in door and the starboard engine's nozzle actuators. The two actuator supply systems (main and back-up) power the aileron, rudder and tailplane actuators; in addition to the engine-driven pump, the port system features an NS-3 autonomous emergency pump ensuring that the aircraft remains controllable in the event of a dual

Tumanskiy R11F2SU-300 engines removed from a Su-15. Note the convergent/divergent nozzle petals and one of the nozzle actuation rams.

Fire suppression system: Titanium firewalls and heat shields are provided in the engine bays to contain possible fires. The Su-15 has an SSP-2I fire warning system (*sistema signalizahtsii pozhahra*; the I means *istrebitel'nyy variahnt* – fighter version), a UBSh-6-1 spherical fire extinguisher bottle and two distribution manifolds around the engines.

Armament: The standard armament of the Su-15 *sans suffixe* consists of two R-98 medium-range air-to-air missiles carried on pylon-mounted BD3-59FK launch rails; the missile comes in SARH (R-98R) and IR-homing (R-98T) versions. The older R-8MR and R-8MT can also be used, but only when the target is attacked in pursuit mode. The missiles are fired singly or in a salvo with a 0.5-second interval.

The Su-15TM is armed with two R-98MR or R-98MT medium-range AAMs and two R-60 short-range IR-homing AAMs carried on pylons installed inboard of the pylons for the larger missiles; later the earlier single-seat versions underwent a similar upgrade, being retrofitted with two extra pylons. The Su-15UM can carry R-98MT and R-60 AAMs for training purposes.

The two fuselage hardpoints can be used for carrying free-fall bombs (Su-15TM only) or UPK-23-250 gun pods (each containing a Gryazev/Shipoonov GSh-23 double-barrelled 23-mm (.90 calibre) cannon with 250 rounds) for use against ground or aerial targets.

Avionics and equipment

Navigation and piloting equipment: KSI-5 compass system, RSP-6 instrument landing system including an ARK-10 ADF, an RV-UM low-range radio altimeter, an MRP-56P marker beacon receiver and SOD-57M DME. An RSBN-5S short-range radio navigation system and an SAU-58 (or SAU-58-2) automatic flight control system are fitted.

Communications equipment: RSIU-5V (R-802) two-way VHF radio (or an R-832M Evkalipt radio on the Su-15TM). Two-seat versions have an SPU-9 intercom (*samolyotnoye peregovornoye oostroystvo*). Single-seat versions have a data link receiver making up part of the Lazoor'-M (ARL-SM) ground controlled intercept (GCI) system.

Weapons control system: The Su-15 is equipped with an RP-15 Oryol-D58 or RP-15M Oryol-D58M fire control radar in a conical radome. The Su-15T features a Taifoon radar in an identical radome, while the definitive production-standard Su-15TM has an RP-26 Taifoon-M radar in an ogival radome. No radar is fitted to the trainer versions. A K-10T collimator sight is provided.

Flight instrumentation: KUSI-2500 airspeed indicator, VDI-30 altimeter, AGD-1 artificial horizon, EUP-53 turn and bank indicator,

Above: The PU-2-8 missile launch rail on the starboard outer wing pylon of a late-production Su-15TM; it is characterised by an unbroken leading edge.

Above: Earlier aircraft were fitted with PU-1-8 launch rails featuring a kinked leading edge.

Above: An APU-60 launch rain on the inboard missile pylon of Su-15TM. These are used for carrying R-60 short-range AAMs.

A UPK-23-250 cannon pod containing a GSh-23 twin-barrel cannon. Two such pods could be carried by the Su-15.

Above: A display mock-up of an R-60 'dogfight missile', showing the canard control surfaces, the so-called 'rollerons' on the wing trailing edges and the protective cover over the IR seeker head.

AM-10 accelerometer (G load indicator), VAR-300 vertical speed indicator (variometr), UKL-2 heading indicator, KI-13 compass, M-2,5 Mach meter, ARK-10 ADF indicator, critical angle of attack warning system and AChKh clock.

Air data is provided by main and back-up PVD-7 pitots; the main pitot is located at the tip of the radome.

IFF equipment: SRZO-2M Kremniy-2M IFF transponder.

Electronic support measures equipment: Sirena-2 or (Su-15TM) SPO-10 Sirena-3 radar warning receiver (single-seat versions only).

Data recording equipment: From Batch 11 onwards, the Su-15 is equipped with an SARPP-12V-1 flight data recorder (*sistema avtomaticheskoy reghistrahtsii parahmetrov polyota* – automatic flight parameter recording system) continuously recording six analogue parameters (barometric altitude, indicated airspeed, vertical G load, tailplane deflection and engine speeds), as well as 12 single actions (gear/flap transition etc.), on photo film. The Su-15UM has an MS-61B cockpit voice recorder.

Rescue system: Sukhoi KS-4 ejection seat permitting safe ejection throughout the flight envelope and on the ground at speeds not less than 140 km/h (87 mph).

Below, left and right: Two views of the instrument panel of a Su-15.

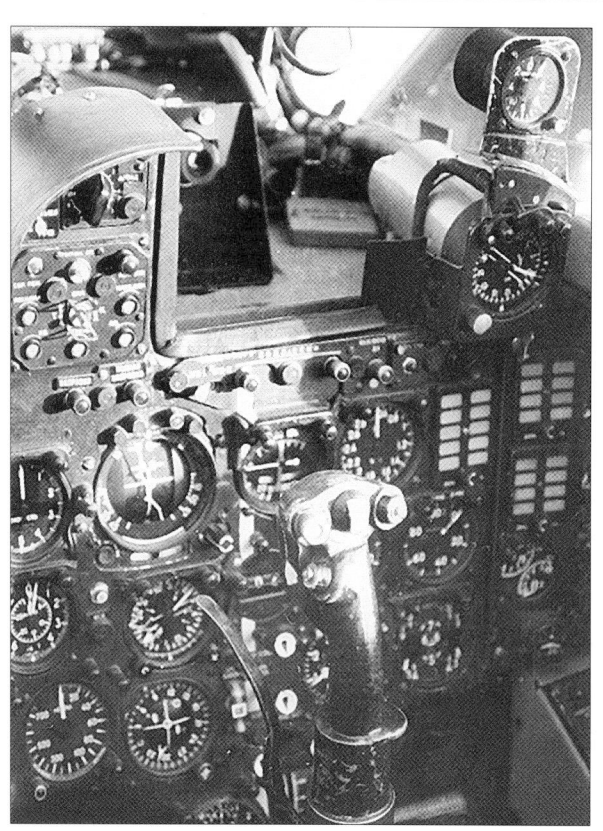

Su-15 specifications

	Su-15	Su-15TM	Su-15UT	Su-15UM
Powerplant	2 x R11F2S-300	2 x R13-300	2 x R11F2S-300	2 x R13-300
Take-off rating, kgp (lbst):				
dry	2 x 3,900 (2 x 8,600)	2 x 4,100 (2 x 9,040)	2 x 3,900 (2 x 8,600)	2 x 4,100 (2 x 9,040)
reheat	2 x 6,200 (2 x 13,670)	2 x 6,600 (2 x 14,550)	2 x 6,200 (2 x 13,670)	2 x 6,600 (2 x 14,550)
Length (less pitot)	20.54 m (67 ft 4⅝ in)	20.54 m (67 ft 4⅝ in)	20.99 m (68 ft 10⅜ in)	19.66 m (64 ft 6 in)
Wing span	8.616 m (28 ft 3¼ in)	9.34 m (30 ft 7¼ in)	8.616 m (28 ft 3¼ in)	9.34 m (30 ft 7¼ in)
Height on ground	5.0 m (16 ft 4⅞ in)	4.843 m (15 ft 10⅝ in)	5.0 m (16 ft 4⅞ in)	4.843 m (15 ft 10⅝ in)
Landing gear track	4.79 m (15 ft 8½ in)	4.79 m (15 ft 8½ in)	4.79 m (15 ft 8½ in)	4.79 m (15 ft 8½ in)
Landing gear wheelbase	5.887 m (19 ft 3¾ in)	5.942 m (19 ft 6 in)	n.a.	5.942 m (19 ft 6 in)
Wing area, m^2 (sq.ft)	34.56 (371.6)	36.6 (393.54)	34.56 (371.6)	36.6 (393.54)
Aileron area, m^2 (sq.ft)	1.126 (12.1)	1.51 (16.23)	1.126 (12.1)	1.51 (16.23)
Aileron balance area, m^2 (sq.ft)	0.34 (3.65)	0.326 (3.5)	0.34 (3.65)	0.326 (3.5)
Horizontal tail area, m^2 (sq.ft)	5.58 (60.0)	5.58 (60.0)	5.58 (60.0)	6.43 (69.1)
Vertical tail area, m^2 (sq.ft)	6.951 (74.74)	6.951 (74.74)	6.951 (74.74)	6.951 (74.74)
TOW, kg (lb):				
normal	16,520 (36,420) †	17,194 (37,905) ¶	16,690 (36,795) §	17,200 (37,920) †
maximum	17,094 (37,685) *	17,900 (39,460) ‡	17,200 (37,920) *	17,900 (39,460) ‡
Empty weight, kg (lb)	10,220 (22,530)	10,874 (23,970)	10,750 (23,700)	10,635 (23,445)
Landing weight, kg (lb)	12,040 (50,790)	12,060 (26,590)	n.a.	13,314 (29,350)
Fuel load, kg (lb)	5,600 (12,345)	5,550 (12,235)	5,010 (11,045)	5,550 (12,235)
Thrust/weight ratio	0.92	0.92	0.88	n.a.
Top speed, km/h (mph):				
at sea level	1,200 (745)	1,300 (807)	1,200 (745)	1,250 (776)
at high altitude	2,230 (1,385) at 15,000 m (49,210 ft)	2,230 (1,385) at 13,000 m (42,650 ft)	1,850 (1,150) at 15,000 m (49,210 ft)	1,875 (1,164) at 11,500 m (37,730 ft)
Mach number at high altitude	2.13	2.16	1.75 3	1.75 8
Climb time to 16,000 m (52,490 ft), minutes	13	n.a.	12	n.a.
Service ceiling, m (ft)	18,500 (60,695)	18,500 (60,695)	16,700 (54,790)	15,500 (50,850)
G limit	5	5	5	5
Unstick speed, km/h (mph)	395 (245)	370 (230)	n.a.	340-350 (211-217)
Landing speed, km/h (mph)	315 (195)	285-295 (177-183)	330-340 (204-211)	260-280 (161-174)
Range, km (miles):				
on internal fuel	1,270 (790)	1,380 (860)	1,290 (800)	n.a.
with drop tanks	1,550 (960)	1,700 (1,055)	1,700 (1,055)	1,150 (715)
Endurance	1 hr 54 min	n.a.	n.a.	n.a.
Take-off run, m (ft)	1,100 (3,600)	1,000-1,100 (3,280-3,600)	1,200 (3,940)	n.a.
Landing run, m (ft):				
without brake parachute	1,500 (4,920)	1,050-1,150 (3,440-3,770)	n.a.	n.a.
with brake parachute	1,000 (3,280)	850-950 (2,790-3,120)	1,150-1,200 (3,770-3,940)	n.a.
Armament:				
missiles	2 x R-98R/T or 2 x R-8MR/MT or 2 x R-8MR1/MT1	2 x R-98MR/MT or 2 x R-8MR/MT or 2 x R-8MR1/MT1	– or 2 x R-8MT 2 x R-60	2 x R-98MT 2 x R-60
cannons	2 x UPK-23-250 ‖ (4 x 23-mm) with 2 x 250 rounds	2 x UPK-23-250 ‖ (4 x 23-mm) with 2 x 250 rounds	– – –	– – –

* With two drop tanks/no missiles
† With two R-98 missiles
‡ With two drop tanks, two R-98Ms and two R-60s
¶ With two R-98 missiles
§ With two dummy R-98 missiles
‖ The UPK-23-250 houses a GSh-23 twin-barrel cannon with 250 rounds

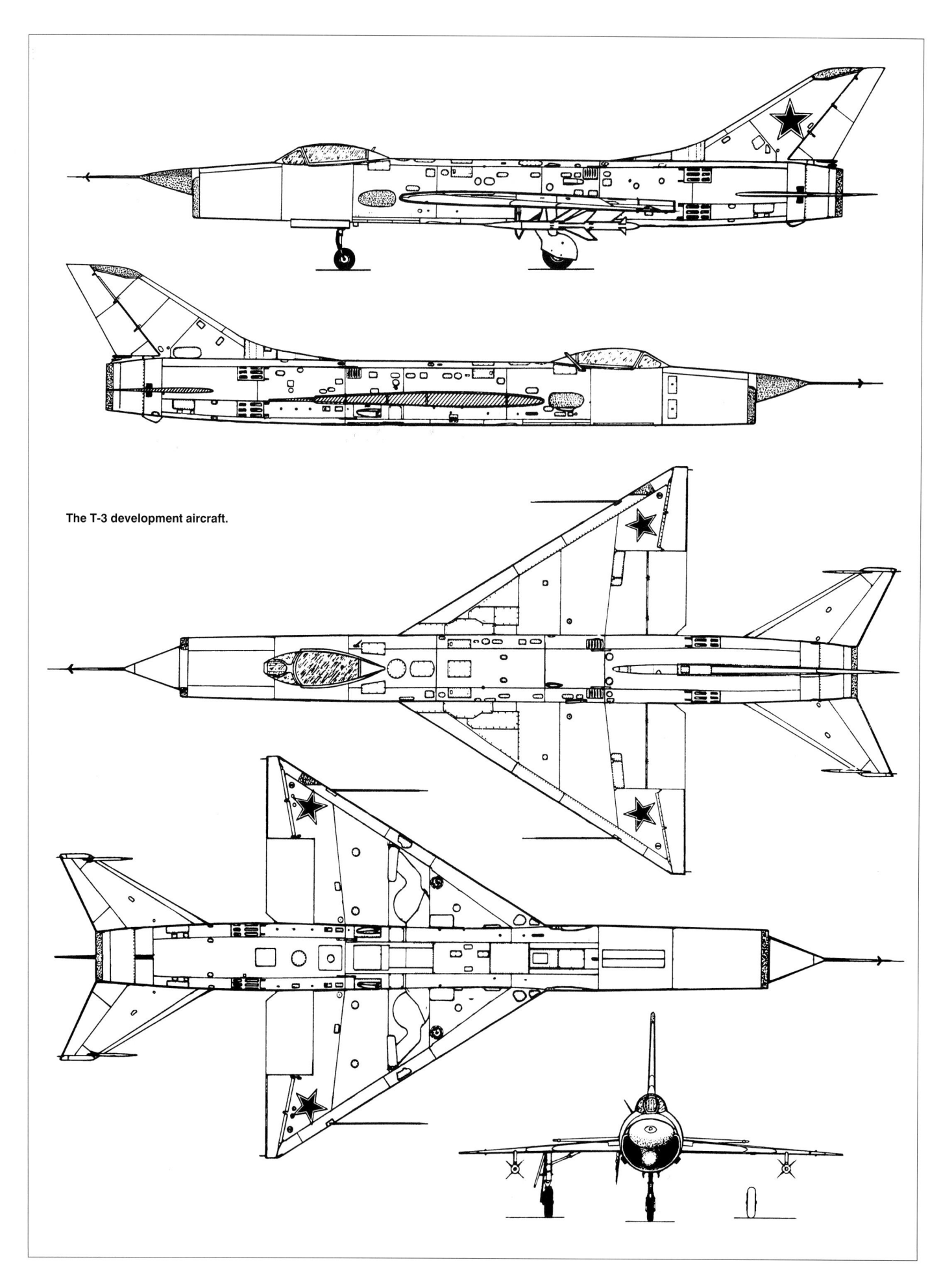

The T-3 development aircraft.

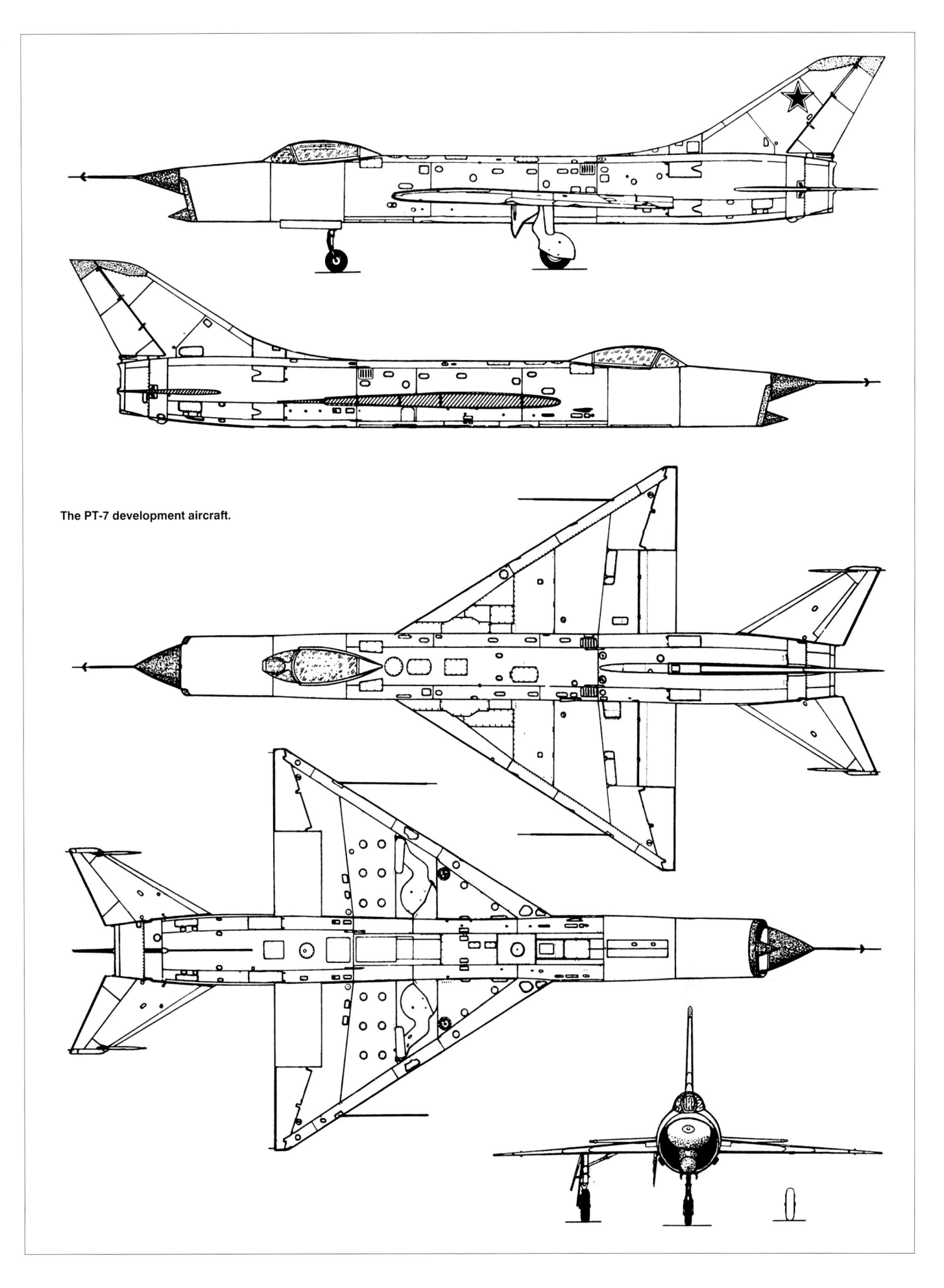

The PT-7 development aircraft.

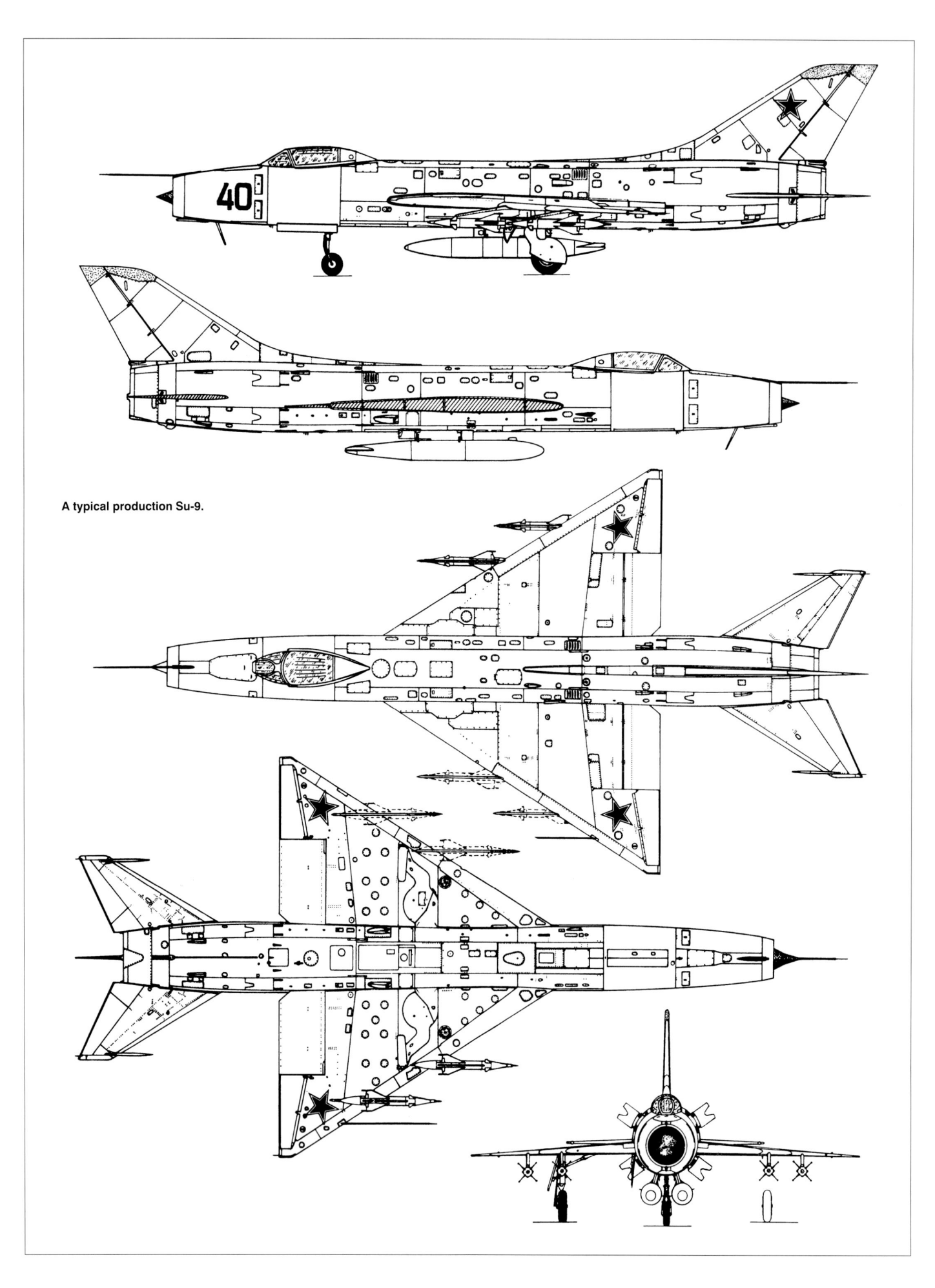

A typical production Su-9.

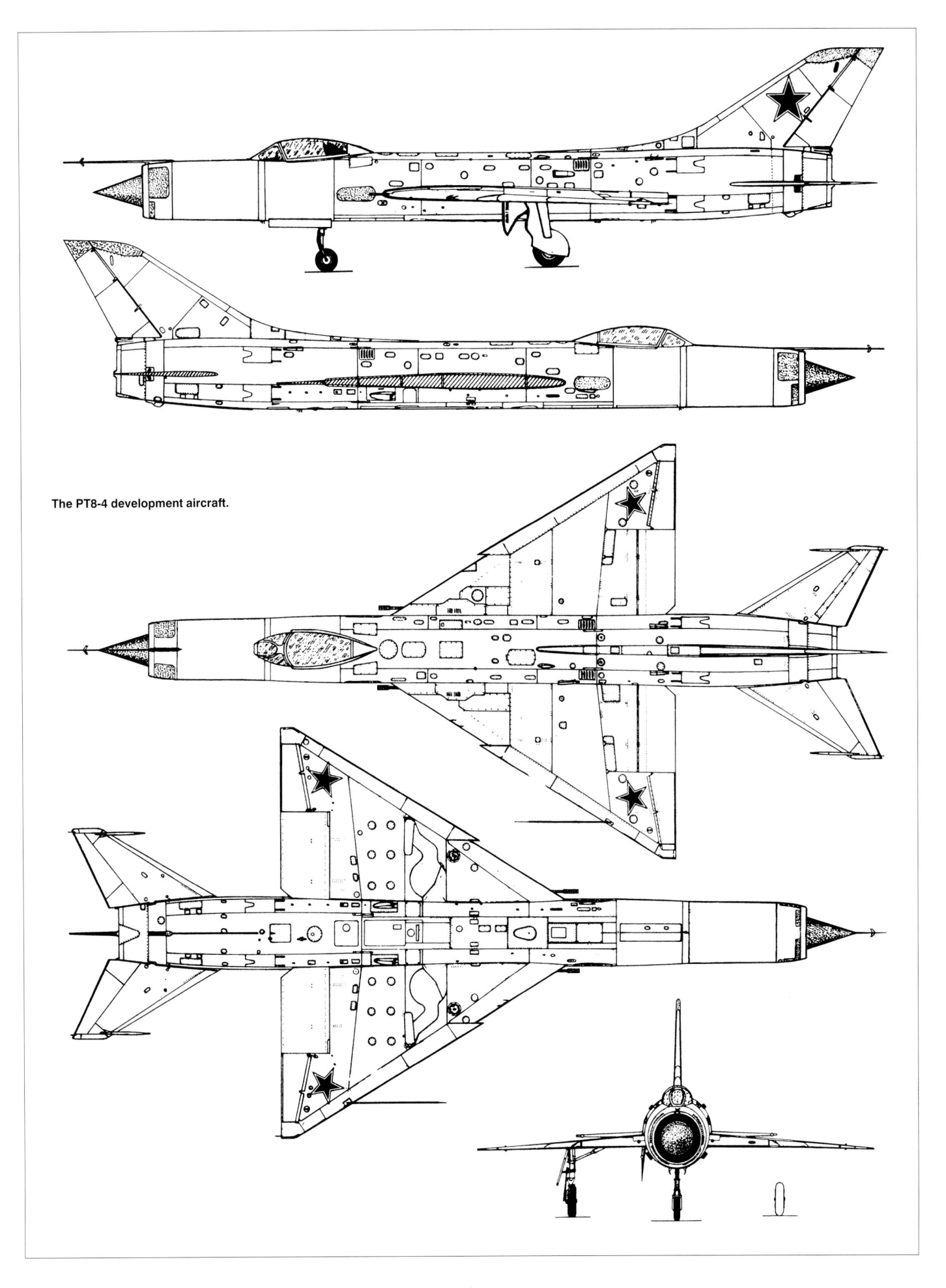

The PT8-4 development aircraft.

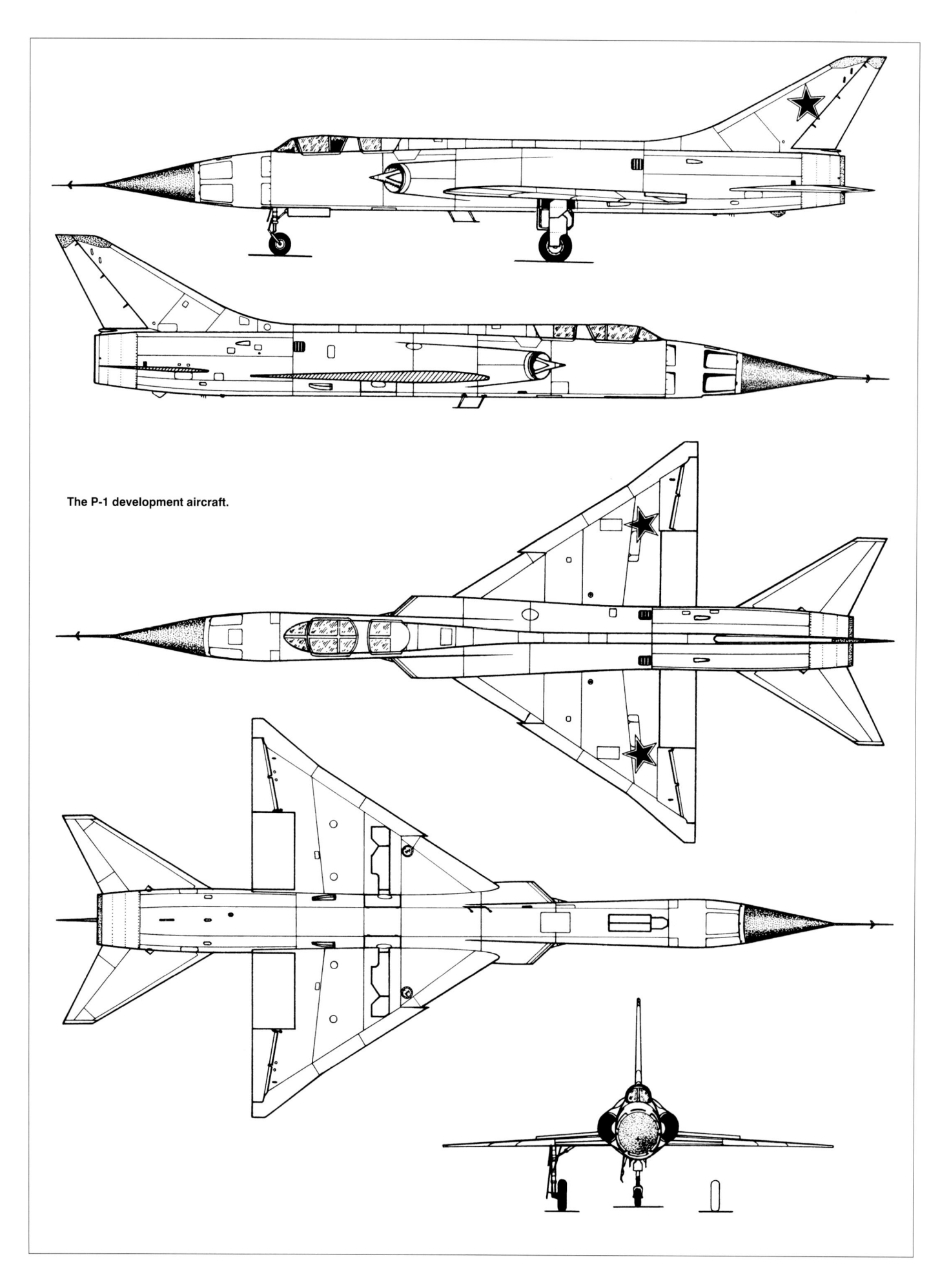

The P-1 development aircraft.

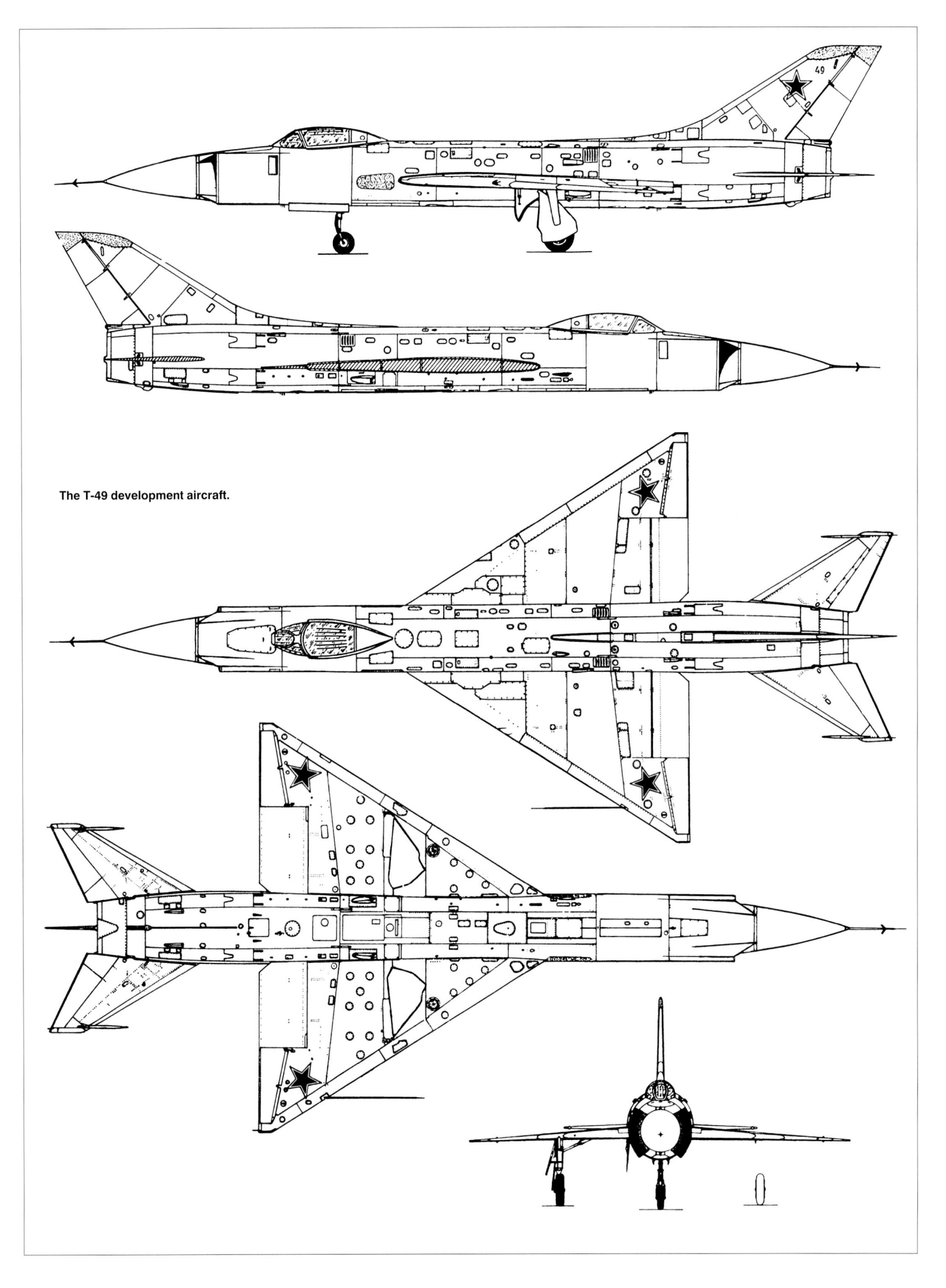

The T-49 development aircraft.

109

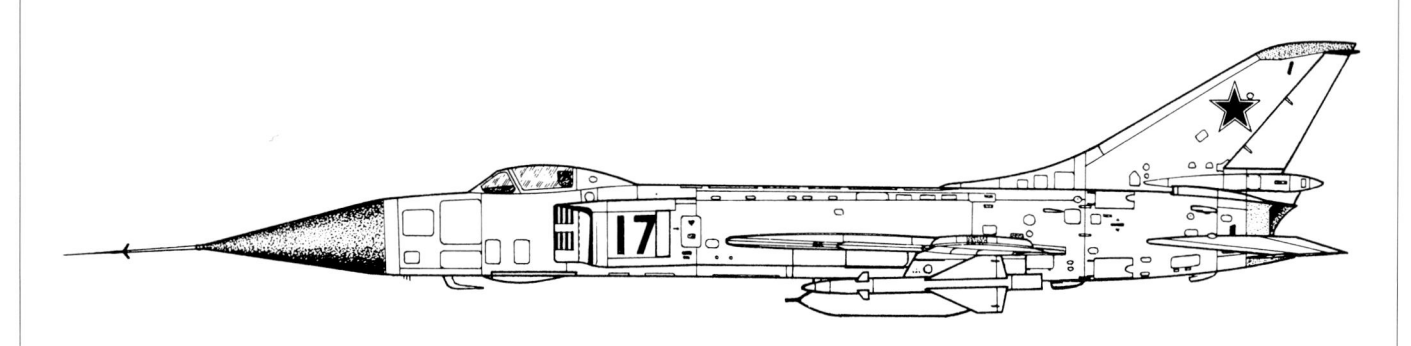

An early-production (pure-delta) Su-15 armed with two R-98 AAMs and two UPK-23-250 cannon pods.

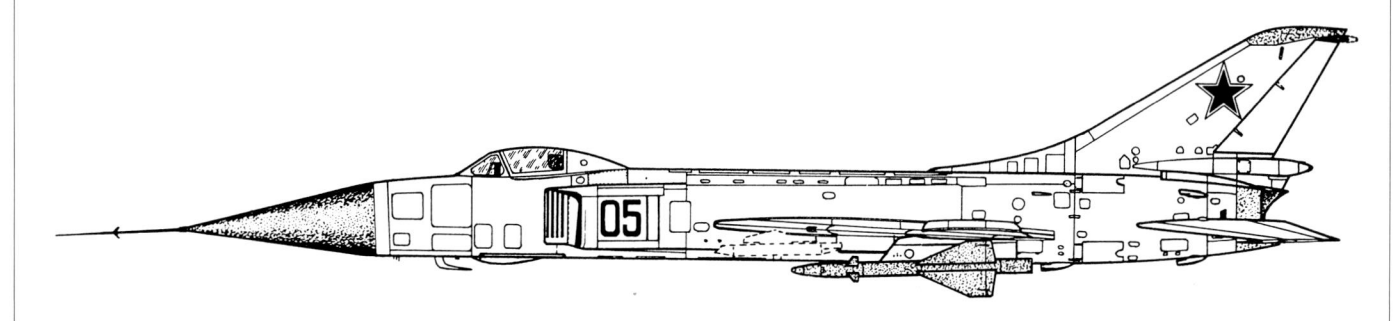

A late-production Su-15 with double-delta wings.

A production Su-15T.

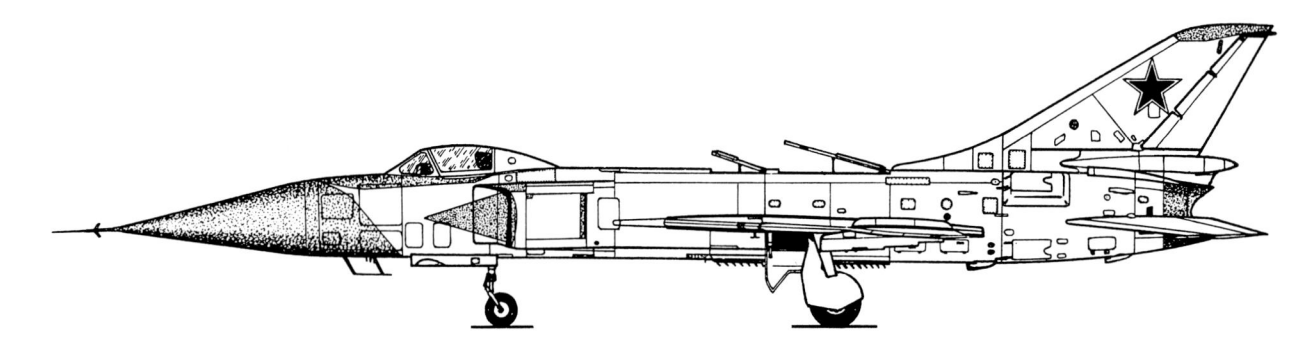

The T-58VD development aircraft with the dorsal air intakes open.

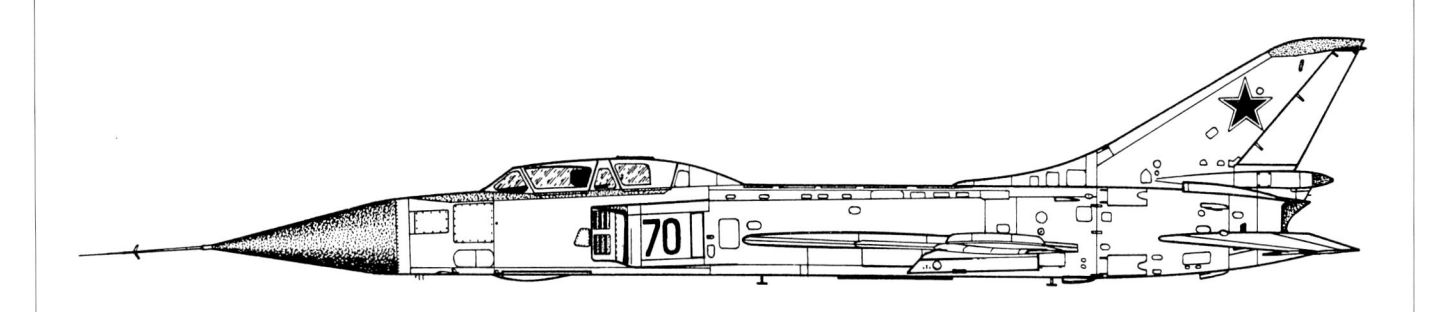

The U-58B development aircraft.

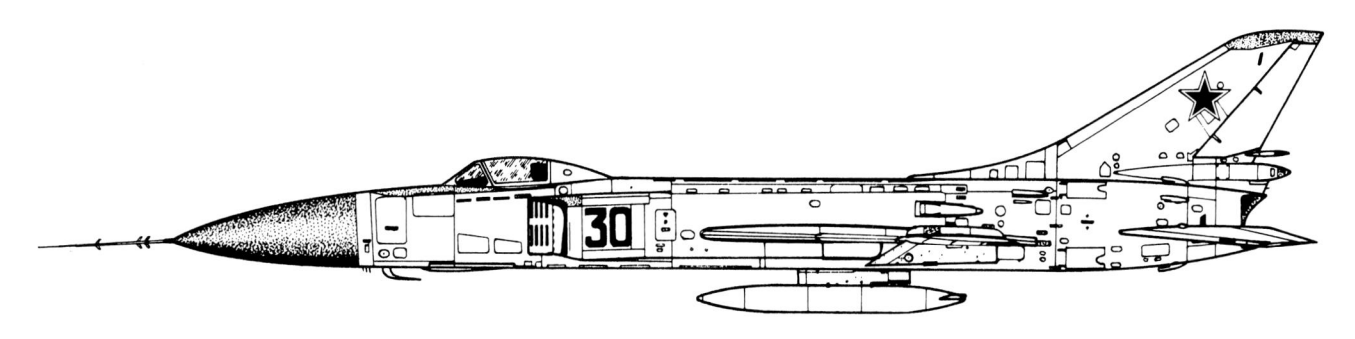

A mid-production Su-15TM with drop tanks.

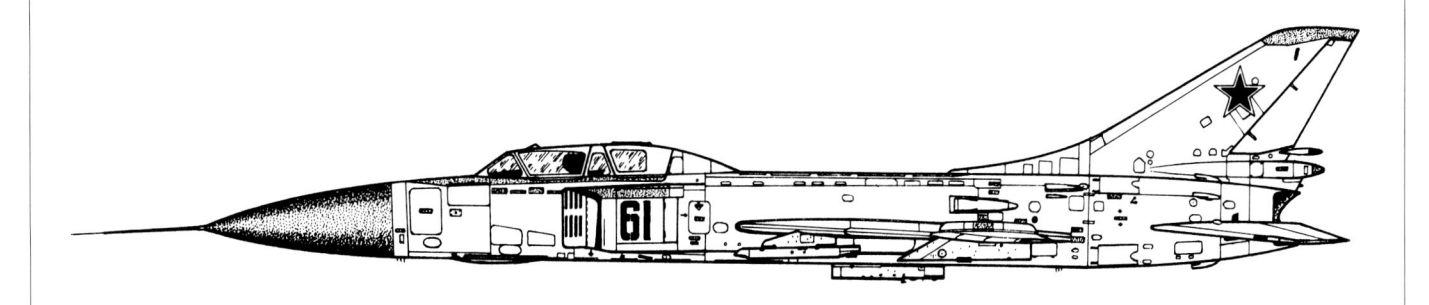

A production Su-15UM trainer.

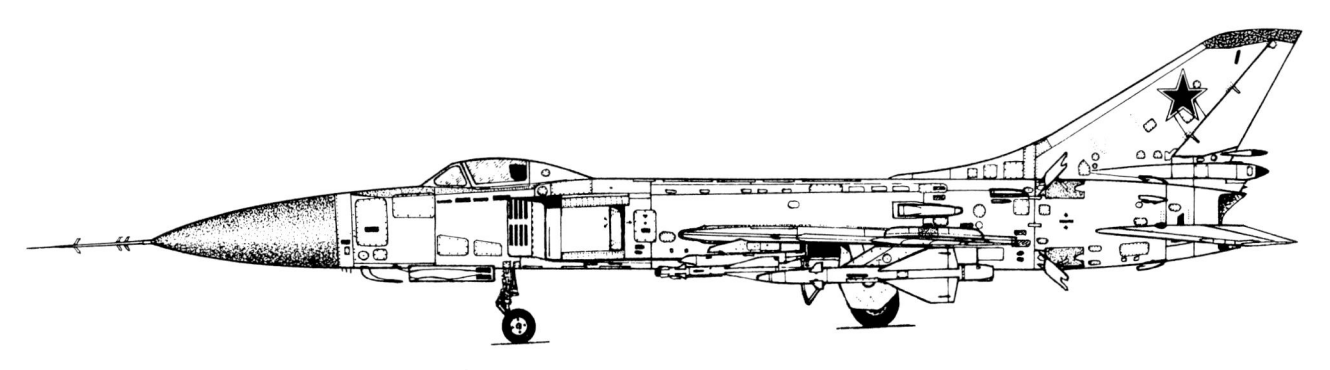

A late-production Su-15TM.

A Su-15T prototype (note the cooling air intakes on the upper rear fuselage).

A three-view of a production Su-15T.

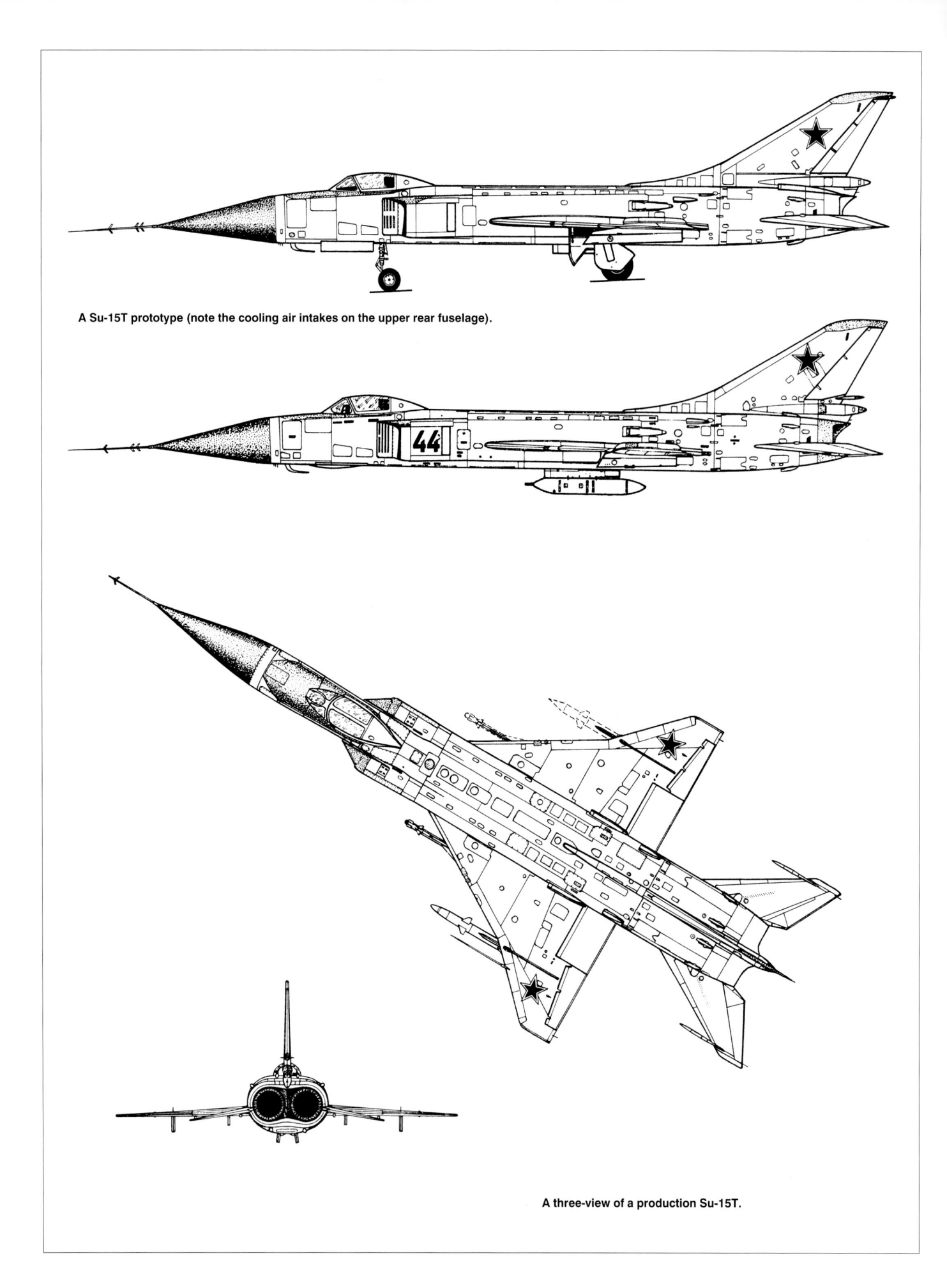

Four views of a late-production Su-15TM armed with two R-98Ms and two R-60s.

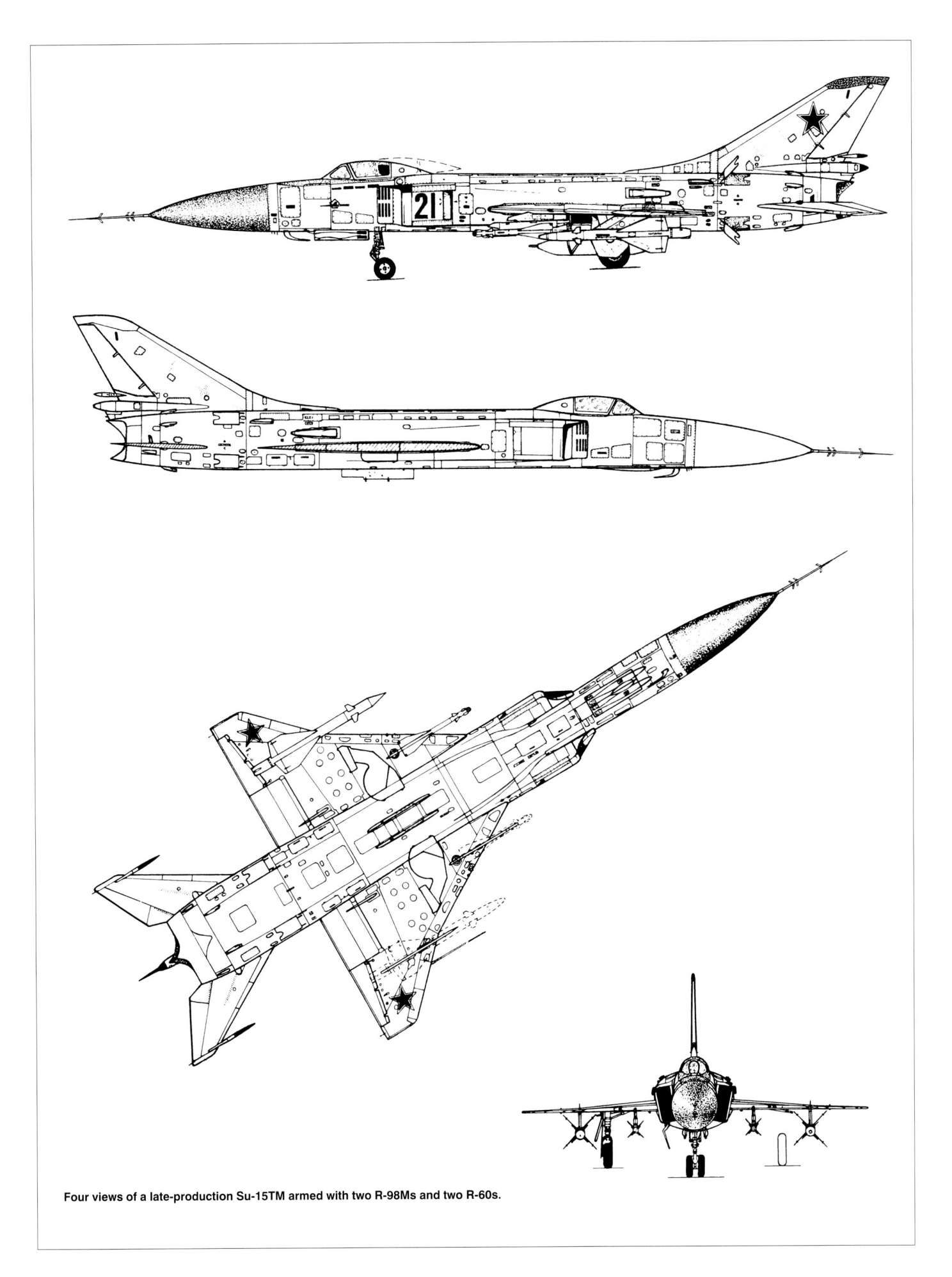

Above: The T43-15 (c/n 1115310), an avionics testbed equipped with the TsD-30TP fire control radar. Note the differently coloured missiles on the inboard and outboard pylons.

The T43-12 development aircraft – the first Su-9 to have an integral No.1 fuel tank – was readily identifiable by the large teardrop-shaped pod low on the starboard side of the nose, apparently housing a TV camera for recording missile launches. Note the red-painted inert RS-2-US missiles.

Above: The T47-5 development aircraft had no radar, as revealed by the all-metal shock cone in lieu of a radome, but could still perform missile launches, as indicated by the nose markings. Here it is seen with inert K-8M AAMs.

The tenth production Su-11 ('10 Blue', c/n 0115310) with two R-8MT missiles.

Above: An atmospheric twilight shot of a late-production Su-9 (note the lack of gun blast plates) on quick-reaction alert duty. The lamps on the landing gear struts confirm to the ground crew that the landing gear is down and locked during approach. The red tactical code is unusual.

This Su-9 ('07 Blue', c/n 100000510) is preserved in the base museum of the PVO's 148th Combat & Conversion Training Centre at Savostleyka AB. Note that the c/n is stencilled at the base of the fin on Moscow-built examples.

Above: Dwarfed by the prototype of the mighty Tupolev Tu-114 Rossiya airliner, Su-11 '14 Blue' (c/n 0115307) sits in the open-air display at the Central Russian Air Force Museum at Monino. The position of the exhibits on the field is regularly changed.

Another view of the Su-11 at Monino accentuating the weathered condition of the aircraft.

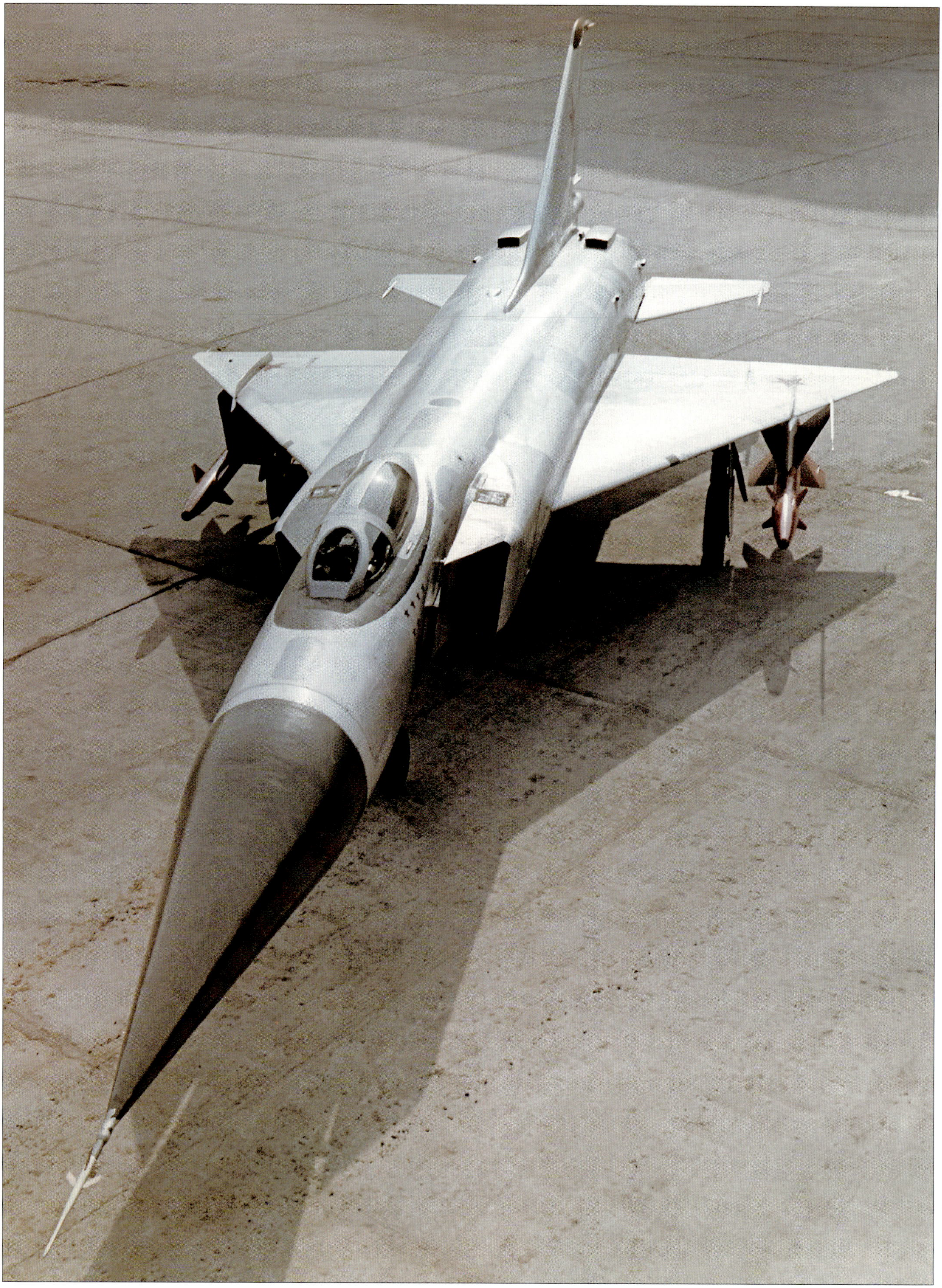

Left, above and below: Three views of the third prototype Su-15 (T58D-3) as originally flown, with an area-ruled centre fuselage and an undernose antenna pod. The aircraft carries dummy K-98 missiles in these photos. The large pod of the radar warning receiver antenna at the tip of the fin was reduced in size on subsequent aircraft.

Above and below: The first Novosibirsk-built pre-production Su-15 (c/n 0015301) as originally flown with the tactical code '34 Red'.

Above and below: '01 Red', the Su-15UT prototype (U58T-1). Though depicted with R-98T missiles, the Su-15UT could not use them for want of missile control equipment.

Above: '05 Blue', a late-production Su-15TM with four missile pylons, configured with two drop tanks.

Su-15TM '23 Red', a camouflaged Tactical Aviation example, sits in storage in company with several sister ships, including another camouflaged 'TM.

Above: Su-15 '42 Red' (c/n 0615342) on display at the open-air aviation museum at Moscow-Khodynka with the radome removed, laying the antenna dish of the Oryol-D58 radar wide open for inspection.

Two Su-15UMs, with '28 Red' foremost, languish at a storage depot, with an Antonov An-12 transport and a Su-15 fighter-bomber in the background.

Above: Su-15TM '38 Red' shares the ramp at the Sukhoi OKB's flight test facility in Zhukovskiy with the P-42 record-breaking aircraft (a highly modified Su-27 fighter) and a Su-24M bomber.

A dramatic shot of a Su-15TM flying a late afternoon practice mission.

Above: Su-15TM coded '11 Blue' peels off, banking away from the camera ship.

Su-15UM '30 Blue' alongside two single-seat Su-15s at the base museum at Savostleyka AB.

Above: Although the Su-15 was not exported, it did wear markings other than Soviet (Russian) red stars. The Ukraine retained a handful of Su-15s after the collapse of the Soviet Union, keeping them operational for a few years.

Another Ukrainian AF Su-15TM. The shield-and-trident insignia on this aircraft is more apparent than on the machine in the upper photo, with traces of the Soviet star still visible underneath.

Above: The Ukrainian Air Force also operated number of Su-15UMs, including '60 Blue'.

The same aircraft undergoes pre-flight servicing in company with a Su-15TM.

We hope you enjoyed
this book . . .

Midland Publishing titles are edited
and designed by an experienced and
enthusiastic team of specialists.

We always welcome ideas from authors
or readers for books they would like to
see published.

In addition, our associate, Midland
Counties Publications, offers an
exceptionally wide range of aviation,
military, naval and transport books and
videos for sale by mail-order worldwide.

For a copy of the appropriate catalogue,
or to order further copies of this book,
and any other Midland Publishing titles,
please write, telephone, fax or e-mail to:

Midland Counties Publications
4 Watling Drive, Hinckley,
Leics, LE10 3EY, England
Tel: (+44) 01455 254 450
Fax: (+44) 01455 233 737
E-mail: midlandbooks@compuserve.com
www.midlandcountiessuperstore.com

US distribution by Specialty Press –
see page 2.

Earlier titles in the series:

Vol.1: Sukhoi S-37 & Mikoyan MFI
Vol.2: Flankers: The New Generation
Vol.3: Polikarpov's I-16 Fighter
Vol.4: Early Soviet Jet Fighters
Vol.5: Yakovlev's Piston-Engined Fighters
Vol.6: Polikarpov's Biplane Fighters
Vol.7: Tupolev Tu-4 Soviet Superfortress
Vol.8: Russia's Ekranoplans

Red Star Volume 9
TUPOLEV Tu-160 BLACKJACK
Russia's Answer to the B-1

Yefim Gordon

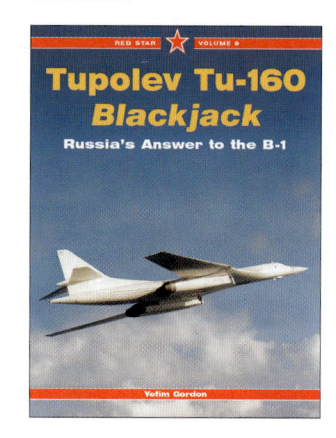

How the Soviet Union's most potent strategic bomber was designed, built and put into service. Comparison is made between the Tu-160 and the Sukhoi T-4 ('aircraft 100', a bomber which was ahead of its time), the variable-geometry 'aircraft 200' – and the Myasishchev M-18 and M-20.
   Included are copies of original factory drawings of the Tu-160, M-18, M-20 and several other intriguing projects. Richly illustrated in colour, many shots taken at Engels.

Sbk, 280 x 215 mm, 128pp, 193 col & b/w photos, dwgs, colour side views
1 85780 147 4   **£18.99**

Red Star Volume 10
LAVOCHKIN'S PISTON-ENGINED FIGHTERS

Yefim Gordon

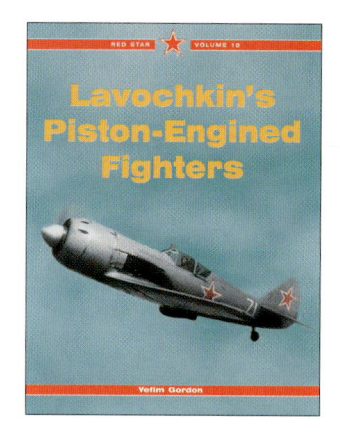

Covers the formation and early years of OKB-301, the design bureau created by Lavochkin, Gorbunov and Goodkov, shortly before the Great Patriotic War.
   It describes all of their piston-engined fighters starting with the LaGG-3 and continues with the legendary La-5 and La-7. Concluding chapters deal with the La-9 and La-11, which saw combat in China and Korea in the 1940/50s.
   Illustrated with numerous rare and previously unpublished photos drawn from Russian military archives.

Sbk, 280 x 215 mm, 144pp, 274 b/w & 10 col photos, 9pp col views, plus dwgs
1 85780 151 2   **£19.99**

Red Star Volume 11
MYASISHCHEV M-4 and 3M
The First Soviet Strategic Jet Bomber

Yefim Gordon

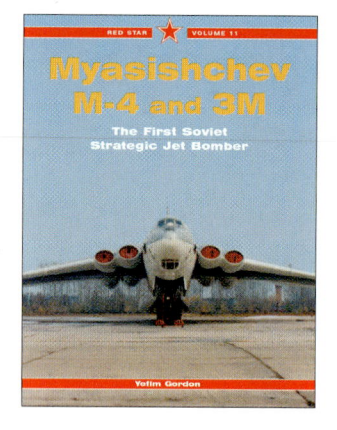

The story of the Soviet Union's first intercontinental jet bomber, the Soviet answer to the Boeing B-52. The new bomber had many innovative features (including a bicycle landing gear) and was created within an unprecedentedly short period of just one year; observers were stunned when the aircraft was formally unveiled at the 1953 May Day parade. The M-4 and the much-improved 3M remained in service for 40 years.

Softback, 280 x 215 mm, 128 pages, 185 b/w, 14pp of colour photographs, plus line drawings
1 85780 152 0   **£18.99**

Red Star Volume 12
ANTONOV'S TURBOPROP TWINS – An-24/26/30/32

Yefim Gordon

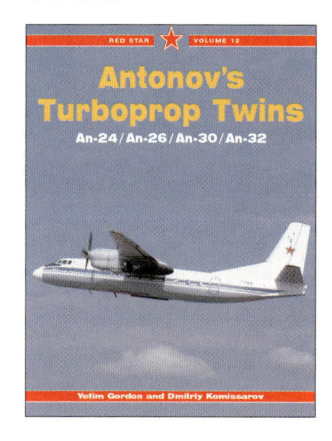

The twin-turboprop An-24 was designed in the late 1950s and was produced by three Soviet aircraft factories; many remain in operation.
   The An-24 airliner evolved first into the 'quick fix' An-24T and then into the An-26. This paved the way for the 'hot and high' An-32 and the 'big head' An-30, the latter for aerial photography.
   This book lists all known operators of Antonov's twin-turboprop family around the world.

Softback, 280 x 215 mm, 128 pages
175 b/w and 28 colour photographs, plus line drawings
1 85780 153 9   **£18.99**

Red Star Volume 13
MIKOYAN'S PISTON-ENGINED FIGHTERS

Yefim Gordon and Keith Dexter

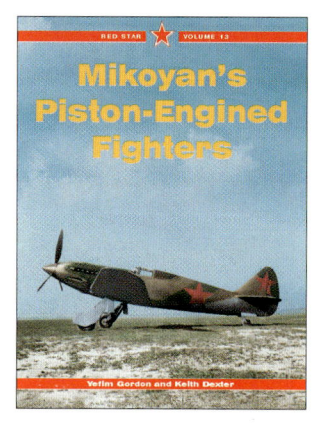

Describes the early history of the famous Mikoyan OKB and the aircraft that were developed. The first was the I-200 of 1940 which entered limited production in 1941 as the MiG-1 and was developed into the MiG-3 high-altitude interceptor. Experimental versions covered include the MiG-9, the I-220/225 series and I-230 series. A separate chapter deals with the I-200 (DIS or MiG-5) long-range heavy escort fighter.

Softback, 280 x 215 mm, 128 pages
195 b/w photos, 6pp of colour artwork, 10pp of line drawings.
1 85780 160 1   **£18.99**

Red Star Volume 14
MIL Mi-8/Mi-17
Rotary-Wing Workhorse and Warhorse

Yefim Gordon and Dmitriy Komissarov

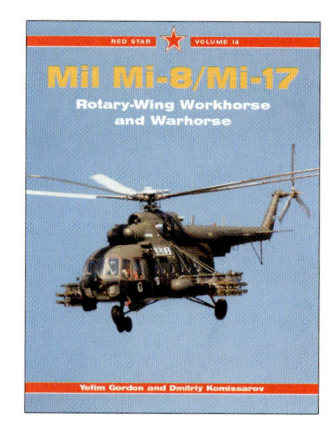

Since 1961, when it first took to the air, the basic design of the Mi-8 has evolved. Every known version, both civil and military, is covered, including electronic warfare, minelaying and minesweeping and SAR. It also served as a basis for the Mi-14 amphibious ASW helicopter.
   Over the years the Mi-8 family have become veritable aerial workhorses, participating in countless wars of varying scale. The type is probably best known for its service in the Afghan War.

Softback, 280 x 215 mm, 128 pages
179 b/w and 32 colour photographs, plus line drawings
1 85780 161 X   **£18.99**

Red Star Volume 15
ANTONOV AN-2
Annushka, Maid of All Work

Yefim Gordon and Dmitriy Komissarov

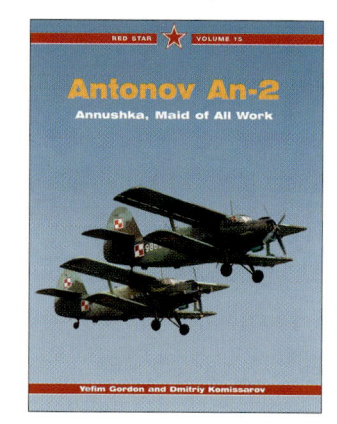

Initially derided as 'obsolete at the moment of birth' due to its biplane layout, this aircraft has put the sceptics to shame. It may lack the glamour of the fast jets, but it has proved itself time and time again as an indispensable and long-serving workhorse. The An-2, which first flew in 1947, has been operated by more than 40 nations.
   The An-2 is the only biplane transport which remained in service long enough to pass into the 21st century!

Softback, 280 x 215 mm, 128 pages
c200 b/w and 28 colour photographs, plus line drawings.
1 85780 162 8   **£18.99**